The Occurrence and Significance of Erosion, Deposition and Flooding in Great Britain

London: HMSO

Acknowledgements

This report was prepared as part of a research project entitled **Review of Erosion, Deposition and Flooding in Great Britain**. The project was funded by the Department of the Environment, under its planning research programme (Contract No. PECD 7/1/412).

This report was written and compiled by Mr E.M. Lee of Rendel Geotechnics, with guidance provided by Dr A. R. Clark (Rendel Geotechnics) and Dr J.C. Doornkamp (Nottingham University). The following sections were based on reports prepared by renowned experts in their fields:

- hillslope erosion and mudfloods (Dr John Boardman, Oxford University);

- river channel instability (Dr Janet Hooke, Portsmouth University);

- river channel sedimentation (Prof John Lewin, University of Wales, Aberystwyth);

- estuaries (Prof John Pethick, Institute of Estuarine and Coastal Studies, Hull University);

Further specialist advice has been provided by:

- Prof Denys Brunsden, Kings College, University of London;

- Prof David Jones, London School of Economics and Political Science;

- Prof Malcolm Newson, Newcastle University.

This review has been undertaken in order to provide **an overview** of erosion, deposition and flooding. It is based on readily available information from major sources and is not a complete review of erosion, deposition and flooding in Great Britain. It is intended to be of use as background information and does not obviate the need for a full investigation of any particular site to establish the suitability for particular types of land use and development.

Executive Summary

Introduction

The operation of physical processes on hillslopes, within river channel networks and at the coast are of considerable interest to landowners, engineers, planners and land managers as well as the scientific community. Erosion processes operate to slowly break down the fabric of the land; the resulting material is then transported by wind or water and deposited elsewhere in new landforms. Land can be lost; channels and reservoirs choked with sediment leading to flood problems or reduced storage capacity; new land can be created; landforms can build up to give increased protection against erosion or flooding.

Erosion, deposition and flooding are natural phenomena and an integral part of the natural landscape, especially the dynamic environments of river channels and the coast. The processes can shape the landscape, forming, for example, the coastal cliffs and broad meandering rivers that are part of our natural heritage. They can create and sustain valued habitats and maintain important recreational beaches or sand dunes.

The processes only become hazards or problems when society encroaches into these dynamic environments either for housing or development on floodplains or coastal cliffs, or for transportation and trade along canals, rivers and estuaries. Here, attitudes to erosion, deposition and flooding can vary dramatically. To some the processes are an acceptable risk associated with living in desirable locations such as close to riverbanks or coastal cliffs; adjustments to the risk, such as floodproofing, can be made to mitigate against the effects of potentially damaging events. Others may be completely unaware of potential problems in an area; to them the sudden occurrence of an event may lead to unacceptable levels of loss.

In other instances the operation of the processes may go largely unnoticed by much of society. Erosion of riverbanks and hillslopes, for example, generally does not involve dramatic events; small amounts of material are regularly detached and carried away. The cumulative effects, however, can lead to serious consequences such as the gradual silting up of canals, rivers and estuaries which make regular dredging essential to maintain navigable watercourses.

This study, commissioned by the Department of the Environment as part of its planning research programme, aims to provide an assessment of erosion, deposition and flooding processes, with particular reference to their significance to land use planning and development. The results have been presented as:

- a review of the occurrence and significance of the processes (this volume);

- a review of approaches to investigation and management of issues related to the processes, with special reference to the role of the planning system (Rendel Geotechnics, 1995);

- a summary report, combining the key components of both the above volumes;

- a computerised database of records from over 1500 significant events over the last 200 years or so;

- a suite of 1:625,000 scale maps of:

  – the distribution of Records of Significant Events,

  – Potentially Vulnerable Areas.

There are a number of key areas where erosion, deposition and flooding has had a significant impact on land use and development:

- slope erosion and mudfloods on hillslopes;

- dust storms on agricultural land;

- flash floods in upland areas;

- bank erosion, sedimentation and channel instability on rivers;

- floods on lowland rivers;

- sedimentation in estuaries;

- floods in low lying coastal areas;

- erosion of coastal cliffs;

- wind blown sand in coastal dunes.

The occurrence and significance of each of these processes is described in this report, with reference to the **impact on the economy** and the **significance for the environment**.

The Impact of Erosion, Deposition and Flooding

That the processes have had notable impacts on the economy is clearly demonstrated by the records of significant events that form the historical database developed as part of this study. These records describe the reported impact of a specific process at a particular location. By collating all the records generated by the same sequence of climatic events, or other causal factors, it has been possible to gain a general indication of the cumulative effects of the different processes operating within the same **erosion, deposition and flooding event**.

A total of 1550 separate events have been identified within the historical sample. The breakdown of event frequency for each decade since 1700 is shown on Figure 1. The pattern reveals an almost exponential increase in frequency of events up to the 1950's, after which the number of events has remained fairly constant. The factors influencing this trend include:

- the rapid spread of development into vulnerable locations throughout the 19th and 20th centuries, resulting in an increase in the risk of a damaging event;

- the significant institutional and structural responses to the major flood disasters of 1947 and 1953 with improved flood warnings and defence schemes based on a better appreciation of catchment and coastal processes, resulting in a reduction of risk in protected areas.

Each event comprises a unique collection of records, each with different affected area, duration and impact characteristics. However, by assessing the cumulative effects of the reported incidents it has been possible to classify each event according to the overall magnitude of the impact. Because of the enormous variety of impacts this has largely been a subjective procedure, although a range of indicative criteria were used to guide the classification process (Table 1). The pattern of events of different magnitude reveals the following average frequencies per decade of this century:

- **Class 1 events** 40 per decade;
- **Class 2 events** 25 per decade;
- **Class 3 events** 7 per decade;
- **Class 4 events** 1 per decade.

It is important to stress that many of the more serious problems have had some form of treatment to reduce the impact of future events of similar magnitude on the community. For example, since the devastating east coast floods of 1953 there has been a major programme of sea defence and flood warning improvements which prevented a repetition of the disaster when an even higher storm surge occurred on 11–12 January 1978.

The problems associated with these processes are generally similar: loss of agricultural productivity; damage to property; the need for maintenance to ensure unhindered use of waterways; and occasionally death or injury. Indeed, it has been estimated from the historical record that the average level of damage or maintenance and defence needs associated with these processes probably exceeds £300M per year (Table 2); these costs are, of course, spread through many levels of the economy, from individuals to industry, local authorities to national government and include:

(i)   **direct damages** caused by the effects of erosion and deposition or the physical contact of floodwater with properties and their contents;

(ii)   **indirect damages** arising as a consequence of direct damage, including: traffic

Figure 1 The frequency of events of different magnitude, from 1700 to 1993

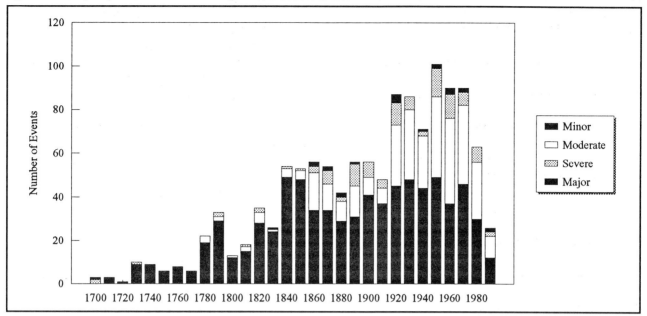

disruption, loss of production, evacuation costs, etc;

(iii)   **intangible damages** ranging from anxiety and stress to ill health related to the general inconvenience caused by the event.

It is, however, the geographical extent and the intensity of damage, disruption or personal losses that set some processes apart. Flooding –flash floods in upland areas, lowland river floods and tidal floods – is the most dramatic and costly problem for society. Examples of particularly distressing and costly events can be found in the historical record for many parts of Britain; some of the worst include:

● the flash floods in the Lynmouth area of Devon on 15 August 1952 when 34 were killed and £9M of damage caused (at 1952 prices);

● the lowland floods of March 1947 which affected rivers throughout South Wales and much of England. The resulting damages were probably in excess of £500M (at current prices);

● the east coast floods of January and February 1953 when over 300 died and damages were estimated at £900M (at current prices);

● the Severnside coast floods of 1606 when about 2000 people were drowned as sea defences were overtopped.

By contrast with the spatially extensive problems associated with flooding, other processes tend to create **site specific** or localised difficulties. Even so, they can still pose a significant threat to construction and development, or lead to high maintenance costs to alleviate the effects of the processes. The erosion of coastal cliffs and sedimentation in rivers or estuaries can be a significant constraint to human activity, as illustrated by:

● the major coastal landslide at Holbeck Hall, Scarborough in June 1993 which is likely to have resulted in excess of £3M of damage and repair works;

● the loss of 75Mm^3 (800 ha) of land from the rapidly erosing Holderness cliffs over the last 100 years;

● annual maintenance dredging costs in excess of £1M are incurred at Harwich and Liverpool. At Kings Lynn, the approaches have to be resurveyed every two weeks with navigation buoys repositioned up to 100 times a year.

Deposition within river channels can lead to serious maintenance and operation problems. In 1993, for example, British Waterways spent over £3M on dredging, involving the removal of 300,000 tonnes of material. These operations are necessary to ensure that British Waterways fulfils its statutory obligations, but the need to dispose of the material on land can lead to conflict with environmental interests.

Table 1 Indicative criteria for the classification of significant events.

MAGNITUDE OF IMPACT	INDICATIVE CRITERIA	
	LOCALISED	WIDESPREAD
Minor event Class 1	Individual towns and villages suffer flooding with no more than ten houses inundated and four houses destroyed, flooding of minor roads. Localised erosion including damage to bridges.	National flooding of agricultural land and infrastructure; regional traffic disruption. Few communities affected. Less than £0.5M damage in total.
Moderate event Class 2	Intense, localised damage in towns or villages; may involve up to five dead. Considerable local disruption, with financial hardships to few.	A region's towns and villages suffer flooding with no more than 2000 houses flooded. Damage is not intense or widespread within a community. Less than £5M damage in total.
Severe event Class 3	Considerable localised damage in towns and villages; may involve up to 15 dead. Event may involve lengthy period of inundation and severe damage to individual properties, widespread evacuation and emergency relief.	A region may experience setbacks to industry, financial hardship to a few and financial setbacks to thousands. Cities may be severely inundated in a number of districts; local towns and villages badly affected. Up to 6000 houses flooded, with damage up to £50M.
Major event Class 4	Almost complete desolation and destruction to a community; may involve up to 30 dead and widespread destruction of property. Evacuation and disaster relief aid required.	Considerable damage to region's cities, towns and villages; may involve over 30 dead. Widespread financial setbacks and financial hardship to thousands of people. More than 10,000 made temporarily homeless, with damages in excess of £50M.

Other processes, such as hillslope erosion, wind erosion and channel migration, can lead to notable problems for the affected landowners and can lead to difficulties where infrastructure and services cross vulnerable areas. The implications of these problems are easy to dismiss as trivial; the following examples should serve to demonstrate that they can lead to serious problems:

● soil erosion and mudflood problems in the South Downs during October 1987 probably resulted in £0.75M of damage, especially in and around Rottingdean;

● channel scour and erosion around the piers of a railway bridge at Glanrhyd, Dyfed led to the bridge collapsing under the weight of a train in October 1987 when four people died;

● the Culbin Sands disaster of 1694, and the following years, led to over 20–30km^2 of fertile farmlands, near Findhorn on the Moray Firth, being buried by up to 30m of loose sand. The estimated damages were probably the equivalent of £25M, at present prices.

Many relatively minor erosion events can also achieve importance because they supply sediment to rivers or the coastal zone. Sediment supply occurs in areas of **hillslope erosion** and where **river channel migration** cuts through floodplain sediments. Once in the channel, the sediment size is important how far it is carried before being temporarily stored in features such as point bars or as spreads on the river bed. In short rivers the suspended load may reach the estuary in a single flood, but coarser sediments may become incorporated in the floodplain. Deposition within the channel can, of course, reduce its capacity and lead to flood problems, as reported for the River Spey in the Grampians and the Findhorn in the Highlands. It can also lead to navigability and loss of reservoir capacity and a decline in water quality. On the coast, erosion can supply sediment to sustain features such as beaches, sand dunes, mudflats and saltmarshes.

The Significance for the Environment

Erosion, deposition and flooding are natural responses to sequences of climatic events. They should not be regarded as wholly detrimental to society's interests as they are necessary for creating

Table 2 An indication of the general order of costs per year arising as a consequence of erosion, deposition and flooding.

	Estimated Annual Cost (£)	Source
Hillslopes	Unknown	
Rivers ● NRA maintenance ● British Waterways dredging	£42M £ 3M	NRA Corporate Plan (1993/94) British Waterways News (1993)
Coast and Estuaries ● NRA maintenance ● dredged spoil	£15M £43M	NRA Corporate Plan (1993/94) Table 8.2; assumed £1 per tonne for disposal
Flood and Coastal Defences ● NRA ● Local Authorities	£115M £65M	NRA Corporate Plan (1993/94) DoE Statistics
Damages ● impact of events in historic record	£50M	Based on values in Table 2.2
Estimated Total	£333M	

Note: Investment in capital works is justified on the basis that its cost is less than the damage avoided, so that without this expenditure the overall cost would be significantly higher.

and maintaining many elements of the natural landscape that we value highly. From a conservation perspective these processes can have important roles in:

● **maintaining** habitats in river environments and on the coast, through regular inundation or supply of sediment;

● **maintaining** geological exposures or valued landforms along the coastline through continued erosion;

● **preparing** gravel bed rivers for spawning fish such as salmon;

● **creating** valued landforms such as the fluvial geomorphological features associated with flash flooding or channel migration;

● **stimulating change** through promoting instability, ensuring that habitats evolve through natural successions, rather than remaining static.

The value of our rivers and coasts is very diverse, ranging from the tourism and recreation importance of features such as sand dunes, beaches and navigable waterways to the scientific and educational benefits of river washlands, coastal cliffs and mudflats. Here, the Government's environmental strategy emphasises the importance of **stewardship** which must underlie the use of environmental resources; balancing the need for

economic growth and prosperity with the safeguarding of the natural world (DoE, 1990a). In many instances this "**sustainable development**" will involve working with natural processes rather than opposing them through engineering works that sacrifice environmental value for a less hazardous setting. The importance of natural processes in now widely appreciated by conservation groups and those bodies with environmental duties (e.g. the NRA). Indeed, this view is central to English Nature's Campaign for a Living Coast (English Nature, 1992).

Erosion and deposition can also play an important role in minimising the impact of extreme events such as floods, by building up "**natural defences**". For example, many coastal landforms offer a degree of protection against coastal flooding. **Sand dunes** serve as a natural barrier against high water levels; this has long been an effective coastal defence for many communities around the coast. **Beaches** and **shingle ridges** absorb as much as 90% of the wave energy arriving at the coast by continuously adjusting their form, providing an important component of sea defences either alone or where they front embankments or sea walls. **Saltmarshes** and **mudflats** are also effective in dissipating wave energy. All these landforms are dependent upon a continued supply of sediment to maintain their form, usually from eroding cliffs; disruption of sediment transport can, therefore, lead to an increase in the degree of risk behind the sediment–starved landforms elsewhere on the coast.

The Occurrence of Erosion, Deposition and Flooding

Erosion, deposition and flooding are associated with hillslopes, rivers and the coast. The **hillslopes** are the site of soil erosion and deposition by water and, occasionally, wing; slope failure through landslides and debris flows also occur (Geomorphological Services Limited, 1986–1987). Both supply sediment to the **river networks** where channel migration, flooding and deposition are the principal forms of landscape activity. The rivers reach the coastline in their **tidal estuaries** where sedimentation, flooding and erosion of soft materials by tidal currents are the dominant processes. On the **open coast** there are complex patterns of erosion and deposition, with tidal flooding in low lying areas.

The landscape within which these processes occur are largely predetermined by topography, the underlying geology and soils, and the land use and vegetation cover. These conditions control the erosion, deposition and flood behaviour of an area or region. For example, catchment characteristics are important controls on the nature and occurrence of floods. However, the geomorphological processes at work on the hills, rivers and coastline are not constant and do not take place in a static landscape. The occurrence of significant events is inexorably linked to the pattern of storms or severe weather conditions that are characteristic of Britain's maritime climate.

The general perception tends to be that erosion, deposition and flooding events are associated with severe rain storms or snow melt conditions. It is important to recognise, however, that the processes can also be the product of an unfavourable combination of land management practice and relatively minor rainfall events (e.g. for hillslope erosion). Erosion and deposition can be a slow progressive process associated with unexceptional seasonal conditions; here, the rate of the process is as much a factor of topography, geology, soil type and land use as climatic events. Significant events may also occur in response to the operation of "normal" physical processes which slowly change the character of a landform until it reaches a critical level at which, for example, minor rainfalls may trigger major events (e.g. sedimentation in rivers can lead to a progressive reduction in channel capacity and, hence, an increase in flood risk). In other instances, accelerated erosion or deposition may be in direct response to the disruption of sediment transport processes, by, for example, constructing a dam across a river or groynes along a coastline.

The Effects of Man

The erosion, deposition and flooding character of many areas has been modified, to varying degrees, by the human occupancy of the landscape and land management practices. Amongst the more significant influences are those associated with: changing agricultural practice, forestry, development, river channelisation and flood defence, coastal defences, dredging operations, water supply and conservation.

In addition to modifying the operation of erosion, deposition and flooding processes, economic growth and residential mobility has resulted in a society that has become more vulnerable to damaging events. This, in part, reflects changing attitudes to natural hazards. For example, in the past, floodplain residents were generally aware of the potential problems and took precautions as and when necessary, moving furniture and valuables upstairs when river levels rose. This long standing voluntary acceptance of the risk has, to an extent, been gradually replaced by a general lack of appreciation of flood problems. High residential mobility in new housing estates has contributed to this situation, as new occupants do not have the accumulated flood experience of residents in more static floodplain communities.

Potential problems have also been heightened by the rise in property values and ownership of expensive household goods such as televisions, washing machines, fitted kitchens, carpets etc., all of which are very vulnerable to flood damage. It has been suggested that between 1977 and 1987 the damage potential rose by up to 100% for long duration flood events. This trend has been reinforced by the continued expansion of housing development in floodplain areas. In Maidenhead, for example, flood damage potential has risen dramatically since the major 1947 event; at that time there were 1400 properties at risk, by 1990 the figure had risen to 3,558; estimated flood damage costs have risen from £1.3M in 1947 to a potential figure of £19M by 1990.

Management of Erosion, Deposition and Flooding Processes

These processes should not be considered in isolation as they are often linked, creating complex problems that can affect many different interests. River erosion may oversteepen slopes and lead to instability; erosion of dunes or mudflats can reduce the effectiveness of "natural" coastal defences and may result in more frequent flooding. In turn, the large volumes of fast flowing water in many floods can cause extensive erosion and, when velocities drop, considerable deposition within a channel, leading to navigation problems.

The effects of erosion, deposition and flooding are complex and should be viewed as expressions of the operation of large physical systems and not as separate processes. Indeed, whilst the effects are readily apparent and well appreciated at a local level, they are not often regarded as the product of the broader controls that influence the behaviour of **river catchments** or **coastal systems**. For example, to understand the flood character of a river, something must be known of the climatic, geological, topographical and land use controls on the supply of water and sediments from the surrounding hillslopes. On the coast, the development of beaches or shingle banks need to be seen as the product of sediment transport within dynamic coastal systems.

The traditional response to erosion, deposition and flooding problems has generally involved river and coastal engineering works to provide protection against potentially damaging events. However, construction of defence works only reduces the risk of damage. **It cannot eliminate the risk**. Increased investment and density of development behind the defences may lead to higher losses when, inevitably, larger events occur.

A balance has to be found between the costs of providing defences and the benefits to the nation as a whole. To attempt to protect all floodplains or the entire coastline would not only be uneconomic but could intensify the problems in many areas. For example:

- construction of flood embankments reduces floodplain storage and can lead to more severe flooding downstream;

- coastal defences may disrupt the supply or transport of sediment around the coast and have an adverse effect on defences elsewhere.

The management of water levels for flood defence or controlling the recession rates of coastal cliffs may also have a significant impact on the environment, particularly wildlife habitats and important geological features.

A growing awareness of the importance of sustainable development, together with greater appreciation of the broad–scale operation of physical processes, have reinforced the need to take a more strategic and pragmatic view of erosion, deposition and flooding problems. This need is currently being expressed through the development of catchment and shoreline management plans which take account of the risks posed by erosion, deposition and flooding whilst recognising their importance to creating or maintaining resources. Traditional hard engineering solutions are being complemented by soft engineering techniques which **work with rather than oppose physical processes**; greater consideration is being given to **avoiding areas at risk** from flooding or erosion, and the need to **minimise the effects of development** on the level of risk elsewhere.

Future Research Needs

In this Report it has been shown how erosion, deposition and flooding can present significant problems for land use planning and development, most notably through:

- the impact of the processes on property, infrastructure and services;

- the effects of development on the degree of risk elsewhere;

- the conflicts generated by the selection of hazard management strategies.

In areas where erosion, deposition and flooding are likely to impose constraints to development and land use, decision makers will need to consider identifying those areas where particular consideration should be given to these issues. This requires access to reliable technical information and although there is considerable volume of available data on erosion, deposition and flooding, little is directly suitable for planners insofar as it could be used without re–interpretation. There is, therefore, a clear need to improve the availability of information that is needed to support forward planning and development control. This could involve:

- providing local authorities with guidance on the use of historical sources for predicting erosion and deposition rates;

- conservation agencies identifying those natural resources that could be affected by a disruption to the natural patterns of erosion, deposition and flooding, and passing this information onto planners and developers;

- encouraging organisations with an interest or responsibility for data collection to develop standard approaches to preparing summary hazard maps and non-technical reports for use by planners and developers;

- developing techniques to evaluate the sensitivity of an area to the effects of development;

- developing an integrated natural hazards databank for use by local planning authorities and developers. This should include information on the nature and distribution of: landslides, undermined ground, natural underground cavities, foundation conditions, natural contamination and erosion, deposition and flooding.

References

Department of the Environment 1990. This Common Inheritance. Britain's Environmental Strategy. HMSO.
English Nature 1992. Campaign for a Living Coast. English Nature, Peterborough.
Geomorphological Services Ltd 1986 – 1987. Review of Landsliding in Great Britain. Reports to the DoE.
Rendel Geotechnics 1995. Investigation and Management of Erosion, Deposition and Flooding in Great Britain. Report to DoE.

Contents

APPENDIX A

List of Figures

List of Tables

LIST OF ABBREVIATIONS

ADAS	Agricultural Development and Advisory Service
AOD	Above Ordnance Datum
AONB	Area of Outstanding Natural Beauty
AoSP	Area of Special Protection (Birds)
BS	British Geological Survey
BSI	British Standards Institution
CCS	Countryside Commission for Scotland
CCW	Countryside Council for Wales
CEC	Crown Estate Commissioners
CMP	Catchment Management Plan
DCPN	Development Control Policy Note
DoE	Department of the Environment
DNH	Department of National Heritage
DSS	Department of Social Services
DTp	Department of Transport
EA	Environmental Assessment
EC	European Community
EN	English Nature
ES	Environmental Statement
GCR	Geological Conservation Review
GDO	Town and Country Planning General Development Order 1988
GVP	Government View Procedure
HWM	High Water Mark
ICE	Institution of Civil Engineers
IDB	Internal Drainage Board
IWEM	Institution of Water and Environmental Management
JNCC	Joint Nature Conservation Committee
LPA	Local Planning Authority
LWM	Low Water Mark
MACC	Military Aid to Civil Communities
MAFF	Ministry of Agriculture, Fisheries and Food
MLWS	Mean Low Water, Spring Tides
MPG	Minerals Planning Guidance note
NERC	Natural Environment Research Council
NCC	Nature Conservancy Council
NCR	Nature Conservation Review
NRA	National Rivers Authority
PDO	Potentially Damaging Operation
POL	Proudman Oceanographic Laboratory
PPG	Planning Policy Guidance note
RIGS	Regionally Important Geological Site
RPA	River Purification Authority
RPB	River Purification Board

RPG	Regional Planning Guidance note
RSPB	Royal Society for Protection of Birds
RTPI	Royal Town Planning Institute
SCOPAC	Standing Conference on Problems Associated with the Coastline
SDD	Scottish Development Department
SMP	Shoreline Management Plan
SNH	Scottish Natural Heritage
SOAFD	Scottish Office Agriculture and Fisheries Department
SPA	Special Protection Area
SSSI	Site of Special Scientific Interest
STWS	Storm Tide Warning Service
UDP	Unitary Development Plan
UK	United Kingdom
WO	Welsh Office
WRVS	Women's Royal Voluntary Service
WWF	World Wide Fund for Nature

1 Introduction

Background

The Department of the Environment (DoE) undertakes geological and earth-science related research as part of its Planning Research Programme. The programme aims to provide information on planning policies, planning processes and the context within which the planning system operates. One of the areas of research is directed towards minerals and land instability (DoE, 1994).

Land which is actually or potentially unstable is widespread in Great Britain. Some instability arises from natural processes. Often it is caused or accelerated by human activities. Problems include landslides and rockfall associated with natural slopes and cliffs, or with embankments, cuttings, quarry faces or spoil tips. Subsidence of the surface is associated with mined ground, natural caves and fissures, solution of salt, or underground combustion of coal. Open, or poorly covered, mine entries may be dangerous. Foundations may be damaged by compression of sediments or of landfill, or by swelling and shrinking of clays or other materials. Britain even experiences damaging earthquakes but, fortunately, these are rare.

In addition, damage can arise from surface flooding from rivers or the sea. Erosion can exacerbate ground movements or cause loss of soils. Deposited sediments can block waterways and pipes leading to flooding problems. Rising groundwater can cause problems for tunnels and basements. Natural chemical substances within the ground may react with buildings materials. Some of these are poisonous to people or livestock. Emissions of natural gases can be explosive, suffocating, poisonous or radioactive.

The costs of these problems in damage to property, injuries and deaths are poorly known but are undoubtedly very large. Many problems can be avoided or reduced by proper precautions. These include planning to avoid inappropriate development in areas most at risk, proper site investigations, appropriate precautionary and remedial works, and design of structures to, for example, accommodate ground movements or exclude hazardous gases. But this is only possible if the problems are appreciated. This requires adequate information for planning of land use, control of development and application of the Building Regulations. Such information can also guide strategies for reclamation of derelict land.

Research is undertaken to establish the nature and extent of problems and the best technical and administrative approaches for dealing with them. The results, thus, contribute to the objectives of the Environment White Paper "This Common Inheritance" of securing a physically safe environment, of minimising risks to human health and the environment, and recycling derelict and contaminated sites (DoE, 1990).

A series of **National Reviews** of specific ground-related problems have been commissioned by the Department:

- Review of Landsliding (Geomorphological Services Ltd 1986-87);

- Review of Mining Instability (Arup Geotechnics, 1991);

- Preliminary Assessment of Seismic Risk (Ove Arup and Partners, 1993);

- Review of Natural Underground Cavities (Applied Geology (Central) Ltd, 1994);

- Review of Foundation Conditions (Wimpey Environmental Ltd, 1994);

- Review of Natural Contamination (British Geological Survey, 1994).

This draft report presents the results of the DoE research contract PECD 7/1/412 entitled "Review of Erosion, Deposition and Flooding in Great Britain". It presents a general review of the occurrence and significance of problems associated with these three natural processes and their impact on the development and use of land (Table 1.1).

Programme of Work

The results presented in this report are based on a general review of:

- the types, causes and extent of significant occurrences or erosion, deposition and flooding in Great Britain;

- problems resulting from the interactions of these phenomena and their relationship to types of land instability and changes in land use; and

- their significance for conservation and maintenance of natural habitats and systems.

Key papers, theses and other documents have been examined to establish the causes of the processes in terms of the controlling framework and preparatory and triggering factors. Special emphasis has been placed on establishing:

- how the various processes are linked between and within physical systems (i.e. river catchments and coastal systems)

- how and why problems arise, taking into account physical and socio–economic factors; and

- the order of costs resulting from specific events.

The significance of processes for conservation and maintenance of natural habitats and systems has been reviewed in terms of:

- the sensitivity of natural habitats to erosion, deposition and flooding and the conservation value of these habitats;

- the importance of processes in maintaining natural systems, including the adverse and beneficial ecological implications.

Records relating to the impact of significant events have been collected from key sources of information in published and unpublished literature. The events considered encompass those involving personal loss, direct economic consequences, indirect costs, environmental costs or environmental gains. The records are intended to **form a sample of data** illustrating the various problems with a fairly even spread across Great Britain.

The broad pattern of significant erosion, deposition and flood events have been established over the last few centuries, concentrating on the period 1780–1993. Information was collected on a standard proforma from a survey of key published and unpublished sources (see the **Methodology report**; Rendel Geotechnics, 1995b), concentrating on: date and location of the impacts; the cause of the event; the impacts; the responses. Each proforma represents a **summary of information** from a single source, and, hence, may contain information about impacts at many sites on the same date or different dates for one site (Figure 1.1).

Over 5000 individual records were collected of locations where erosion, deposition and flood processes have resulted in significant losses. These 5000 records correspond with 1500 separate events (defined as a collection of **related** impact records occurring on the same day or over a similar period). Thus, records of flood damage in Nottingham, Shrewsbury, York, London, Selby and many other sites in March 1947 are classified as part of the same event.

The information has been stored in a computerised database, developed with dBase IV software. A key feature of the database is its ability to link between records, events, locations, geomorphological systems, processes and dates. The dBase "runtime" facility allows the information to be accessed on any DOS–based personal computer without the need for the user to purchase a copy of the software, and as such is compatible with a vast range of other software and hardware platforms. Further details of the database structure are presented in the **Methodology report** (Rendel Geotechnics, 1995b). The database is held on open file in the Department of the Environment.

Table 1.1 The aims and objectives of the study

The **aim** of the study has been to provide an assessment of the nature of erosion, deposition and flooding processes in Great Britain, and an understanding of their significance for the development and use of land, and for the operation of the planning system.

The **objectives** of the study were to review:

a)   the types, causes and extent of significant occurrences of erosion, deposition and flooding in Great Britain;

b)   problems arising from the interaction of these phenomena and their relationships to types of land instability and changes in land uses;

c)   their significance for conservation and maintenance of natural habitats and systems;

d)   methods for investigating and monitoring erosion, deposition and flooding;

e)   measures for hazard reduction including planning, preventive and remedial techniques;

f)   the current availability and reliability of information needed for formulating planning policies and deciding planning applications;

g)   methods of risk assessment, including the significance of flooding in relation to existing uses of land and land use potential, and the definition of areas in which development is constrained by erosion, deposition or flooding;

h)   methods of summarising and presenting information so that it can be taken into account readily in the planning and development process;

i)   the administrative and legislative framework for dealing with the problems;

j)   gaps in existing knowledge, and to set out priorities for further investigations; and

k)   to prepare a draft framework of advice to planners and developers.

The work has been organised in four related Tasks, as follows:

Task 1:   Occurrence and Significance of Erosion, Deposition and Flooding in Great Britain.

Task 2:   Review of British and overseas practices for investigation, management and risk assessment.

Task 3:   Review of the planning system responses to hazards, and improvements to practice.

Task 4:   National and summary reports and dissemination of results.

This report addresses Task 1: The occurrence and significance of erosion, deposition and flooding.

The record is a **sample** not a systematic listing of all significant events to date. This can be illustrated with reference to Figure 1.2 which compares the flood history of the Severn at Bewdley, as recorded by a gauging station, with the sample of significant events incorporated into the database. In general, the information sources from which the sample has been derived are biased towards the newsworthy; this can, of course, vary from day to day and over the decades. The information presented in these sources may not be completely reliable and can be biased towards certain types of damage or loss. The sources concentrate on the impact on built up areas; similar or even larger events in rural areas may have been ignored by those compiling the historical sources. The database is, therefore, a record of the nature and location of significant impacts that have arisen because of erosion, deposition and flooding processes over the last 200

years or so. It is not necessarily, however, a reliable record of where major events have occurred, all the affected locations or where they could occur in the future.

The frequency of recorded events for each decade since 1700 is shown on Figure 1.3, which reveals a dramatic and steady increase throughout this period. The underlying reasons for this trend are likely to be many and varied. In part it reflects the spread of development into vulnerable locations, although this needs to be offset by the possible increase in media interest in the effects of natural hazards. It is clear, however, that despite the extensive structural and institutional responses to erosion and flooding over the last 50 years, there has not been a significant reduction in the number of damaging events that occur in each decade.

Figure 1.1 The historical records database: events, records and proformas.

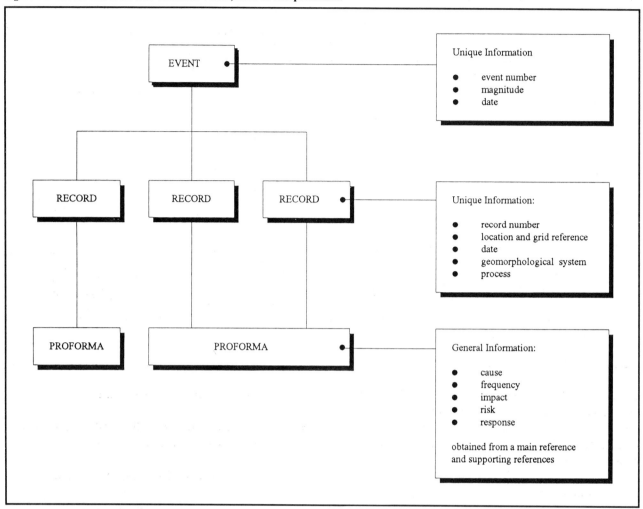

The distribution of records by county or Scottish administrative region is presented in Table 1.2 and indicates a variable distribution with highest densities in the following areas:

- Mid Glamorgan (17.9 records/10km^2)
- Greater London (16.2 records/10km^2)
- Isle of Wight (10.5 records/10km^2)
- Gwent (10.1 records/10km^2)

Both the distribution map and density figures (Figure 1.4 and Table 1.2, respectively) reveal marked concentrations in the Thames Valley, South Wales and the central valley of Scotland, which broadly matches the pattern of population density. There are, however, some areas of unexpectedly low reported concentrations including: Merseyside and Greater Manchester, the central south coast of England, Humberside and the Lake District. The patchiness of the distribution therefore raises questions as to the extent to which these concentrations reflect the true incidence of damaging events as against spatially variable reporting. It is almost certain that the sample has

under-represented events in some areas, whereas in other the number of reported incidents may over-state the relative significance.

Limitations

The aim has been to produce a **national overview** of the problems associated with erosion, deposition and flooding not a detailed appraisal of specific problems in individual areas. The limitations, therefore, should be self-evident. The report should not be relied upon as a source of detailed information for supporting planning or development decision-making; where detail is presented it is largely illustrative and should not be taken as indicative of a systematic detailed treatment of all problem areas.

The report aims to highlight those areas where particular problems may occur, it would therefore be the responsibility of local authorities and developers to seek further advice on the nature and

Figure 1.2 Flood history of the River Severn at Bewdley compared with the frequency of historical events.

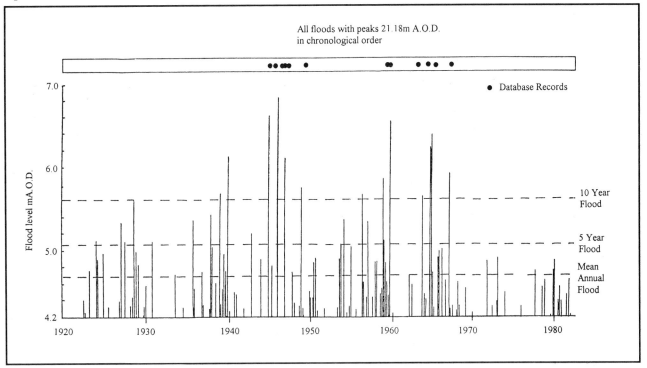

extent of problems experienced at local level. In this context, attention is drawn to a number of key organisations, with important roles in hazard management, who should be consulted when considering specific problems:

i.   **Flood defence** on rivers and the coast;

        the National Rivers Authority
        (NRA; England and Wales)
        Internal Drainage Boards (IDBs;
        England and Wales)
        Local Authorities

ii.   **Coast protection** against the effects of erosion;

        Maritime District councils (England and Wales)
        Island and Regional councils (Scotland)

Guidance on how further detailed investigations can be structured and undertaken is to be included in a companion report entitled "Investigation and Management" (Rendel Geotechnics, 1995a).

Contents of this Report

The results of Task 1 are presented as:

- a written report (this volume);

- a suite of two 1:625,000 scale maps each comprising four sheets:

  – Potentially Vulnerable Areas;

  – Record Locations; presenting the locations of recorded impacts of significant events.

- a methodology report (Rendel Geotechnics, 1995b).

Figure 1.3 The frequency of recorded significant events since 1700.

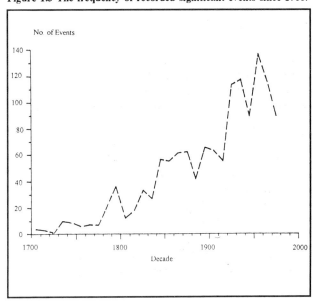

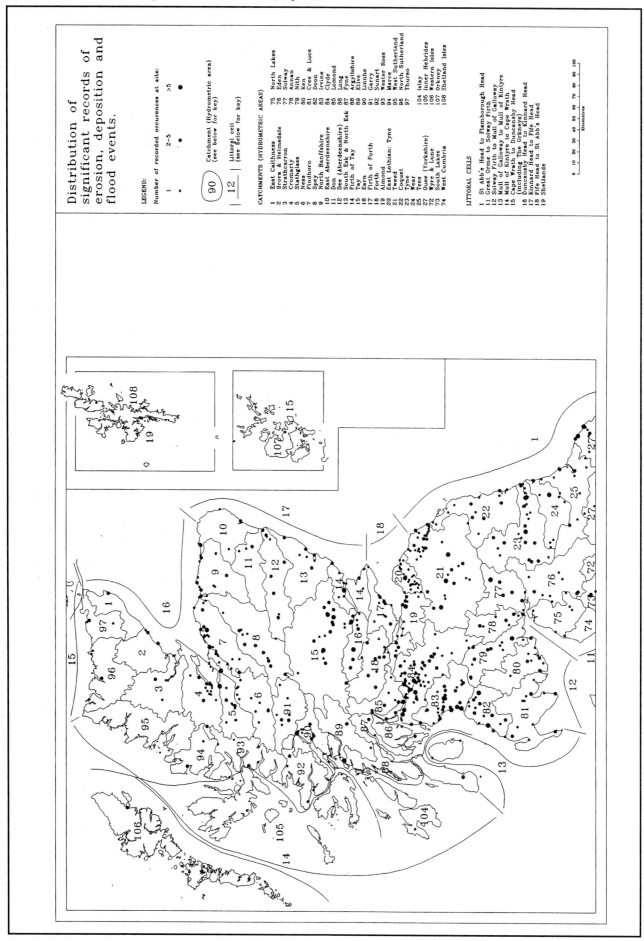

Figure 1.4 (cont ...)

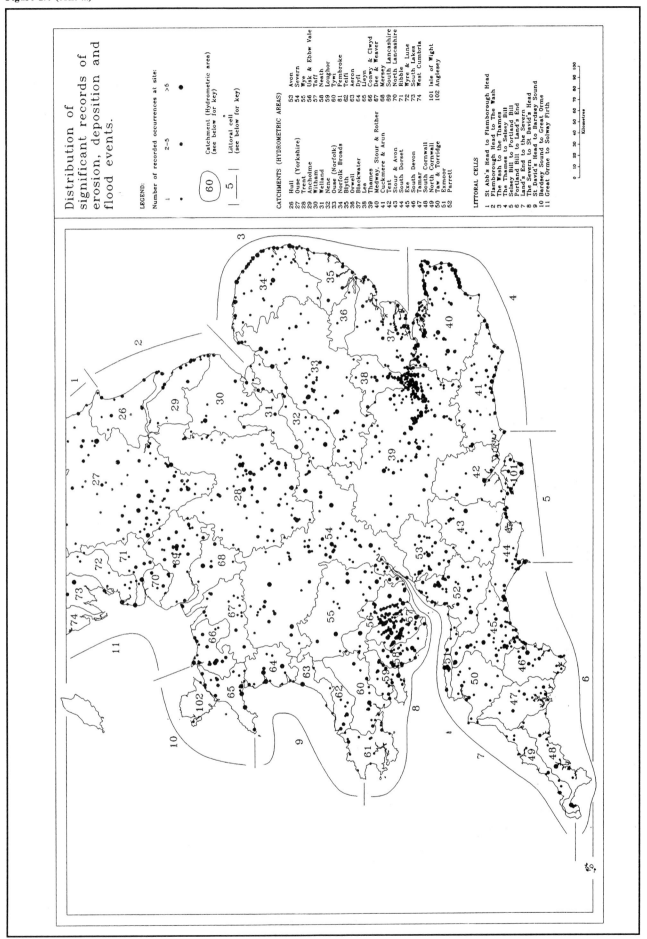

Distribution of significant records of erosion, deposition and flood events.

Table 1.2 The relative frequency and density of events and records by county and Scottish Administrative Region

County	Population	Area (km²)	Pop density (per km²)	Number of: Events	Records	Density (per 100km²) Events	Records
Unspecified	–	–	–	217	308	–	–
Avon	919800	1346	683	41	75	3.05	5.57
Bedfordshire	514200	1235	416	13	20	1.05	1.62
Berkshire	716500	1259	569	47	110	3.73	8.74
Borders	103881	4672	22	27	86	0.58	1.84
Buckinghamshire	619500	1883	329	15	21	0.80	1.12
Cambridgeshire	640700	3409	188	21	53	0.62	1.55
Central	267492	2629	102	25	35	0.95	1.33
Cheshire	937300	2328	403	13	22	0.56	0.95
Cleveland	541100	583	928	5	7	0.86	1.20
Clwyd	401900	2427	166	18	31	0.74	1.28
Cornwall	469300	3548	132	30	62	0.85	1.75
Cumbria	486900	6810	71	24	39	0.35	0.57
Derbyshire	914600	2631	348	24	44	0.91	1.67
Devon	998200	6711	149	87	208	1.30	3.10
Dorset	645200	2654	243	74	106	2.79	3.99
Dumfries & Galloway	147805	6370	23	35	88	0.55	1.38
Durham	589800	2436	242	13	16	0.53	0.66
Dyfed	431600	5758	75	38	85	0.66	1.48
East Sussex	670600	1795	374	35	54	1.95	3.01
Essex	1495600	3672	407	45	97	1.23	2.64
Fife	341199	1307	261	18	26	1.38	1.99
Gloucestershire	520600	2643	197	21	33	0.79	1.25
Grampian	503888	8704	58	42	93	0.48	1.07
Greater London	6377900	1579	4039	97	257	6.14	16.28
Greater Manchester	2454800	1287	1907	19	45	1.48	3.50
Gwent	432300	1376	314	50	139	3.63	10.10
Gwynedd	238600	3870	62	47	119	1.21	3.07
Hampshire	1511900	3777	400	16	33	0.42	0.87
Hereford & Worcester	667800	3927	170	45	88	1.15	2.24
Hertfordshire	951500	1634	582	18	27	1.10	1.65
Highland	204004	26136	8	69	223	0.26	0.85
Humberside	835200	3512	238	29	48	0.83	1.37
Isle of Wight	126600	381	332	32	42	8.40	11.02
Kent	1485100	3731	398	155	304	4.15	8.15
Lancashire	1365100	3063	446	31	62	1.01	2.02
Leicestershire	860500	2553	337	22	53	0.86	2.08
Lincolnshire	573900	5915	97	34	71	0.57	1.20
Lothian	726010	1755	414	24	90	1.37	5.13
Merseyside	1376800	652	2112	7	8	1.07	1.23
Mid Glamorgan	526500	1023	515	55	186	5.38	18.18
Norfolk	736400	5368	137	81	142	1.51	2.65
Northamptonshire	572900	2367	242	13	22	0.55	0.93
Northumberland	300600	5032	60	36	67	0.72	1.33
North Yorkshire	698700	8309	84	58	104	0.70	1.25
Nottinghamshire	980600	2164	453	35	58	1.62	2.68
Orkney	19612	881	22	1	1	0.11	0.11
Oxfordshire	553800	2608	212	30	43	1.15	1.65
Powys	116500	5074	23	37	60	0.73	1.18
Shetland	22522	1427	16	4	4	0.28	0.28
Shropshire	401600	3490	115	35	46	1.00	1.32
Somerset	459100	3451	133	35	128	1.01	3.71

Table 1.2 (cont ...)

County	Population	Area (km^2)	Pop density (per km^2)	Number of: Events	Records	Density (per 100km^2) Events	Records
Somerset	459100	3451	133	35	128	1.01	3.71
South Glamorgan	383300	417	919	23	41	5.52	9.83
South Yorkshire	1248500	1560	800	20	40	1.28	2.56
Staffordshire	1020300	2716	376	14	22	0.52	0.81
Strathclyde	2248706	13851	162	147	373	1.06	2.69
Suffolk	629900	3797	166	30	60	0.79	1.58
Surrey	998000	1679	594	28	55	1.67	3.28
Tayside	383848	7503	51	64	138	0.85	1.84
Tyne & Wear	1087000	540	2013	23	36	4.26	6.67
Warwickshire	477000	1981	241	14	31	0.71	1.56
Western Isles	29600	2898	10	4	4	0.14	0.14
West Glamorgan	357800	817	438	35	63	4.28	7.71
West Midlands	2499300	899	2780	13	23	1.45	2.56
West Sussex	692800	1989	348	13	21	0.65	1.06
West Yorkshire	1984700	2039	973	30	60	1.47	2.94
Wiltshire	553300	3581	155	20	34	0.56	0.95
	54048567	229419	236	2521	5190	1.10	2.26

NOTE: Events are counted no more than once in each county/Scottish region
Population based on the preliminary reports of the 1991 census of England, Wales and Scotland.

Chapter 1 : References

Applied Geology (Central) Limited, 1994.
Review of Natural Underground Cavities. Report to DoE.

Arup Geotechnics, 1991. Review of Mining Instability. Reports to DoE.

British Geological Survey, 1994. Review of Natural Contamination. Reports to DoE.

Department of the Environment 1990. This Common Inheritance. Britain's Environmental Strategy. HMSO.

Department of the Environment, 1994. Geological and Minerals Planning Research Programme. Annual Report for 1993/94.

Geomorphological Services Limited, 1986–1987. Review of Landsliding. Reports to DoE.

Ove Arup and Partners, 1993. Preliminary Assessment of Seismic Risk. Reports to DoE.

Rendel Geotechnics, 1995a Investigation and Management of Erosion, Deposition and Flooding in Great Britain.

Rendel Geotechnics, 1995b Erosion, Deposition and Flooding in Great Britain. Methodology Report. Open File Report held at DoE.

Wimpey Environmental Limited, in prep. Review of Foundation Conditions. Reports to DoE.

2 Background to the Erosion, Deposition and Flood Character of Great Britain

Introduction

The operation of geomorphological processes on hillslopes, within river channel networks and at the coast are of considerable interest to landowners, engineers, planners and land managers as well as the scientific community. Erosion processes operate to slowly break down the fabric of the land; the resulting material is then transported by wind or water and deposited elsewhere in new landforms. Land can be lost; channels and reservoirs choked with sediment leading to flood problems or reduced storage capacity; new land can be created; landforms can build up to give increased protection against erosion or flooding.

Erosion, deposition and flooding are natural phenomena. They are an integral part of the natural landscape, especially the dynamic environments of river channels and the coast. The processes can shape the landscape forming, for example, the coastal cliffs and broad meandering rivers that are part of our natural heritage. They can create and sustain valued habitats and maintain important recreational beaches or sand dunes.

The processes only become hazards or problems when society encroaches into these dynamic environments either for housing or development on, for example, floodplains or coastal cliffs or for transportation and trade along canals, rivers and estuaries. Here, attitudes to erosion, deposition and flooding can vary dramatically. To some the processes are an acceptable risk associated with living in desirable locations such as close to riverbanks or coastal cliffs. Adjustments to the risks, such as floodproofing can be readily made to mitigate against the effects of potentially damaging events. Others may be completely unaware of potential problems in an area; to them the sudden occurrence of an event may lead to unacceptable levels of loss.

In other instances the operation of the processes may go largely unnoticed by much of society. Erosion of riverbanks and hillslopes, for example, generally does not involve dramatic events; small amounts of material are regularly detached and carried away. The cumulative effects, however, can lead to serious consequences such as the gradual silting up of canals, rivers and estuaries which make regular dredging essential to maintain navigable watercourses.

Processes and Mechanics : Erosion

Erosion is a two–stage process comprising **detachment** of material, ranging from individual soil particles on hillslopes to enormous coherent blocks of material in coastal landslides, and the **transport** by water or, less frequently, wind. The severity of erosion depends on the quantity of material supplied by detachment and the ability of the running water or wind to carry it. Without removal, material will accumulate and act as a natural control of erosion. Thus, two significant conditions can be recognised (Morgan, 1980a):

(i)   **supply limited** where more material could be transported by, for example, a river than is supplied by collapse of the riverbanks;

(ii)   **transport limited** where more material is supplied than can be transported, as for many major coastal landslides where debris lobes on the foreshore can take many years to be removed by the sea.

Erosion can be viewed as a function of the power of water or wind (**erosivity**) and the resistance of the material (**erodibility**). The erosivity of rainfall, for example, is measured in terms of the **kinetic energy** associated with particular storms. Maps of rainfall erosivity, such as Figure 2.1, tend to show

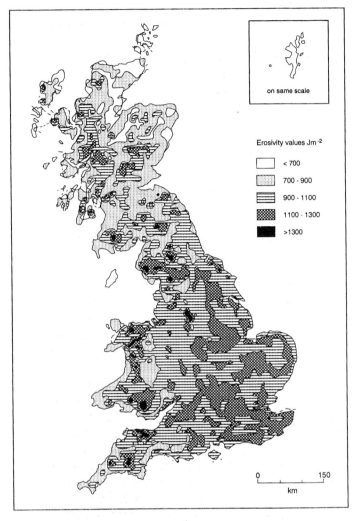

Figure 2.1 Rainfall erosivity in Great Britain (after Morgan, 1980b).

on same scale

Erosivity values Jm⁻²

☐	< 700
▨	700 - 900
▤	900 - 1100
▦	1100 - 1300
■	>1300

0   150
km

Figure 2.2 Critical velocities for erosion, transport and deposition as a function of particle size for water (top) and wind (bottom) (after Morgan, 1980a).

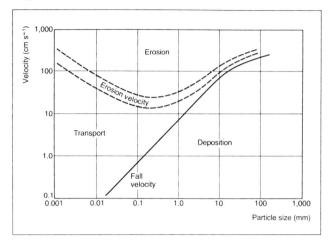

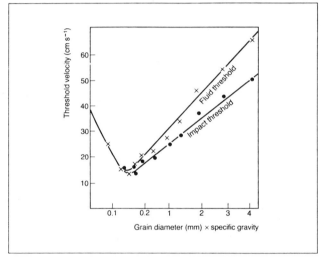

a variable pattern, with the highest values generally associated with upland areas of western Britain. However, when compared with other parts of the world, rainfall erosivity values are very low (Morgan, 1980b).

The erosivity of running water is determined by the **velocity of flow** which, in turn, is a function of the flow depth, channel gradient and the roughness of the stream bed. The water velocity must overcome the resistance of the materials before erosion can occur. The threshold value, or critical velocity, varies with the grain size of the material (Figure 2.2a); generally a larger force is required to move larger particles, although for grains smaller than 0.5mm the critical velocity increases with decreasing grain size because of the cohesion between clay particles.

There are two main components of river bank erosion;

●   on a seasonal basis with, for example, supply through collapse in summer

followed by removal under higher flows in winter;

●   in sequences of floods. The result of one flood can make a channel more vulnerable to the effects of a following flood, for example, by causing incision and exposing more bank material to erosion.

The effectiveness of river flows in achieving erosion can be described in terms of the rate of energy supply at the channel bed i.e. **the stream power per unit area**, (this is a product of the discharge and slope angle divided by the flow width). Figure 2.3 shows the distribution of stream power at bankfull discharge for various British rivers. This reveals a trend for more powerful rivers in the upland areas to the north and west; indeed, there is often a thousand-fold difference between upland and lowland rivers.

On the coast, the potential for erosion is provided by wind-generated waves and tides. The energy

Figure 2.3 Stream power at bankfull discharge (after Lewin, 1981).

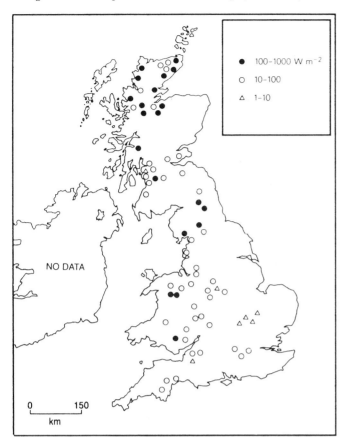

arriving at the coast is many times greater than along even the largest river channels. For example, the South Wales coast receives energy generated over 25M km² of the North Atlantic, whereas the River Severn catchment is only around 25,000km². There is no simple measure of the wave erosivity environment around Great Britain, although extreme wave height (Figure 2.4) highlights the exposed nature of much of western Britain.

The erosive power of wind is a function of its velocity, with critical velocities for picking up and carrying particles varying with grain size in a similar way to running water (Figure 2.2b). Available wind energy and, hence, erosivity is greatest on the north and western coasts (Figure 2.5).

Although resistance to erosion can be determined by slope steepness, topographic position and vegetation cover, the material properties are the most important factor in controlling **erodibility**. Resistance to erosion varies with a wide range of soil and rock properties, most notably soil texture or bedrock lithology, shear strength, water content or hydrogeology and structure. On hillslopes and alluvial river channels, silty and fine sandy soils are most susceptible to erosion to both water and wind, as they are most easily detached and transported (Figure 2.2). Dry soils are more susceptible to wind erosion than moist or wet soils.

The most erodible bedrocks tend to be the sedimentary rocks of Lowland Britain, especially the clay-rich strata (Figure 2.6). A slope is only as strong as its weakest horizon and, thus, the presence of clay strata in coastal cliffs will generally give rise to instability, even when overlain by sandstones and limestones, as on the West Dorset coast (see Chapter 10). The older rocks of Upland Britain, by contrast, are generally much harder, having been toughened and altered by geological processes. They tend to be more resistant to erosion and, hence, form the higher ground.

The overall pattern of a general increase in erodibility towards the south and east of England is complicated by the presence of weak, lithologically variable sediments known as **superficial deposits** (Figure 2.7). These surface sediments represent the debris produced by the operation of geomorphological processes (wind, water, ice, gravity, the sea), mainly during the Quaternary (the last 2.4 million years). It is estimated that thick (i.e. greater than 2m deep) spreads of superficial deposits cover 41% of Great Britain (33% of England, 59% of Scotland and 30% of Wales) and in certain areas, such as East Anglia and North-east England, wholly conceal the underlying bedrock for hundreds of square kilometres with extensive blankets of material up to 70m thick. The nature and composition of these materials is extremely varied, including floodplain alluvium and alluvial lowlands fringing the coast, coastal spits of sand and gravel, river terrace deposits, windblown sand and silt (loess) and even the spreads of debris produced by landslides, especially the widespread downslope sludging of materials that occurred during past cold climate episodes (solifluction). But the overwhelming bulk of superficial deposits are of glacial origin, produced by the ice-sheets that repeatedly expanded over the northern half of Britain (Figure 2.8). These materials were either derived directly from the ice as beds of "boulder clay" (till) with subordinate lenses of sand and gravel, or laid down as spreads of sand and gravel by meltwaters issuing from decaying ice-sheets (outwash or fluvio-glacial deposits).

Superficial deposits are often very erodible. Indeed, the highest rates of coastal recession occur on the boulder clay cliffs of Holderness and East Anglia (Chapter 10). In upland areas, these deposits may also be prone to water erosion, releasing significant

13

Figure 2.4 Extreme wave heights around the British Isles (after NERC, 1991).

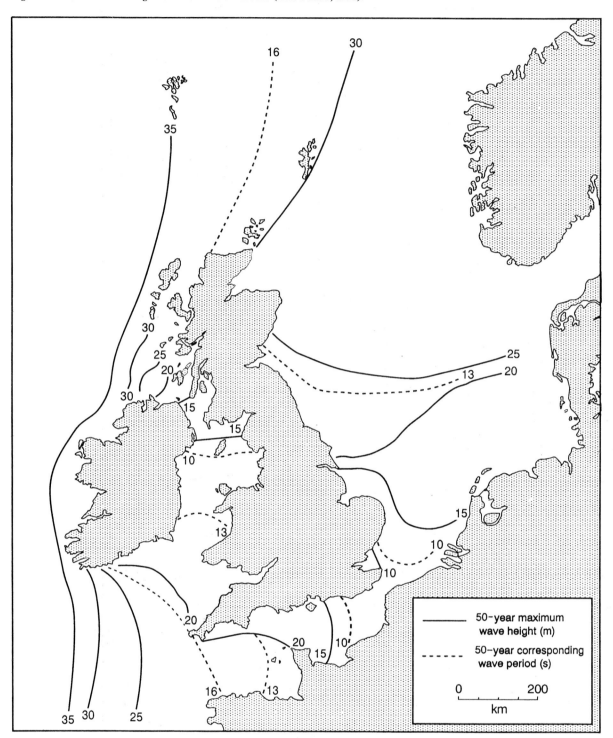

Processes and Mechanisms: Sediment Transport and Deposition

volumes of sand, gravel and cobbles and giving rise to the characteristic gravel–bed rivers of Upland Britain.

Eroded material is carried by transporting agents – wind or water – until the velocity of flow drops below the **fall velocity threshold**, causing the material to be deposited (Figure 2.2). As the fall velocity varies with grain size, an important distinction needs to be made between:

● non cohesive sediments comprising individual grains of **sand** or **gravel**;

● cohesive sediments containing significant amounts of **clay** whose electro–magnetic properties bind individual grains together to

Figure 2.5 Extreme wind speeds around the British Isles; gust speed (in knots) with a recurrence of 50 years (after NERC, 1991).

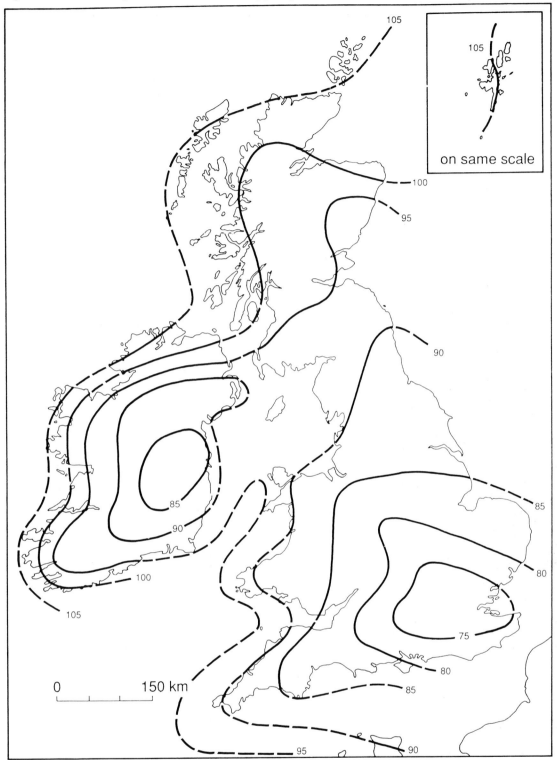

give them a bulk strength or cohesion.

On hillslopes, eroded soil tends to be deposited in fans or spreads at the base of the slope, where the change in slope angle causes a drop in water velocity. Deposition can also occur behind hedgerows and walls which act as barriers to runoff. The amount of sediment that is carried into watercourses can be extremely variable. For example, measurements of soil loss at a variety of

sites in England suggest that between 25–90% of eroded soil may be transported beyond the boundaries of eroding fields (Quine and Walling 1991). Although some may be deposited before reaching a stream, it is clear that loss of finer sediments (fine sands, silts and clays) from. hillslopes may represent an important source of river borne material. Much coarse material (gravel, cobbles and boulders) will only reach a stream in extremely severe events.

Figure 2.6 The geology of Great Britain (after Jones and Lee, 1994)

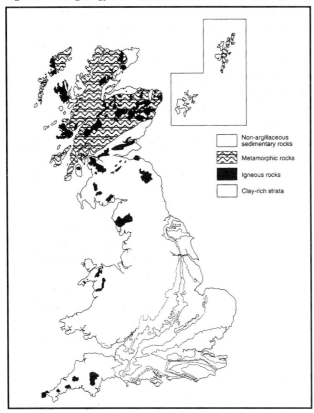

Non-argillaceous
sedimentary rocks

Metamorphic rocks

Igneous rocks

Clay-rich strata

Rivers and streams transport sediment in a variety of ways:

(i)   **in solution**; soluble materials may be dissolved in water and flushed into river systems, especially during the initial stages of a storm. In terms of the total amount of material removed by rivers, the solute component can be high;

(ii)   **in suspension**; suspended sediments are transported in the flowing body of a watercourse where they are supported by turbulent eddies in the flow. The finest material (clays) may be fairly evenly distributed through a stream, whereas coarser materials may be more concentrated near the bed.

Suspended loads are greatest during flood events, although river turbulence is usually adequate to carry fine sediments even at quite low flows.

(iii)   **as bed load**; material can be moved along the stream bed, although the processes are very complex. Indeed, the mixed sizes of bed materials in natural streams tends to complicate the theoretical relationships between grain size and flow (e.g. Figure

2.2); finer materials shelter between coarser particles, coarser material may provide a layer of "armour" on the river bed, protecting the underlying fine materials. It is important to recognise, however, that once the critical threshold velocity has been exceeded, the transport rate varies as the cube of the velocity. Thus small increases in velocity can make an enormous difference to sediment transport. At large floods the whole bed can be mobile.

(iv)   **as buoyant load**; organic debris such as leaves or wood can be carried by the flow. This debris may be stranded or deposited in quiet backwaters or behind structures such as culverts and bridges where they may form temporary dams in periods of high flows.

The amount of material in transport in rivers is closely related to the discharge. However, the overall load of sediment is generally limited by the supply of material from adjoining land or the channel itself, i.e. **supply limited** conditions. In Britain, annual rates of sediment yields from river catchments are typically 100 tonnes/km^2 of less, although higher values have been recorded in upland catchments.

The pathways by which sediment is moved through a river system can be complex. Some fine material may be transferred from an eroding hillslope into the sea in a single flood event; coarse bed material may move only tens of metres from an eroding riverbank to a gravel bar within the channel. Although it is difficult to give precise figures, research has shown that a significant proportion of a rivers sediment yield may not reach the sea directly, being "stored" in a variety of landforms:

(i)   temporary **in-channel accumulations** of generally fine material which is flushed clear by large flows. Surveys of the lower Severn at Tewkesbury, for example, have shown a summer build up of fine material which is subsequently moved downstream in winter;

(ii)   the development of **gravel bars** (shoals) or spreads of fine material in still water, as the channel shifts laterally (see Chapter 6).

(iii)   **channel margin deposits** laid down close to riverbanks during floods. The lower Severn, for example, has prominent natural levees up to 3m higher than the

Figure 2.7 Superficial deposits in Great Britain (after Dearman and Eyles, 1982).

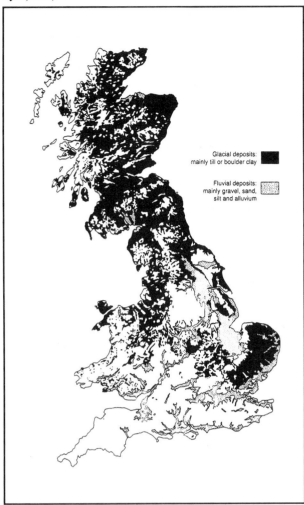

Glacial deposits: mainly till or boulder clay

Fluvial deposits: mainly gravel, sand, silt and alluvium

surrounding land.

(iv) **floodplain deposits** which may be laid down broadly to the limits of flood inundation. In practice, however, higher rates of deposition tend to be localised, as in abandoned channels or in depressions where ponded floodwaters allow sediment to settle from suspension. Research has shown that up to 50% of suspended sediment maybe conveyed out of a channel and onto a floodplain during floods. In exceptional floods, floodplain accretion can exceed 10cm (Newson and Macklin, 1990).

(v) **sedimentation of lakes or reservoirs,** causing a reduction in water storage capacity (see Chapter 3).

(vi) progressive sedimentation of estuaries which are generally major depositional sites for fine sediments (see Chapter 8), and creating large expanses of alluvial soils.

Sediment storage is generally the prevailing condition for most river systems. However, it is important to appreciate the connections between sites of erosion and sediment stores. Prevention of erosion upstream can cause bar erosion or stabilisation downstream; shoal removal upstream can cause bank erosion and subsequent shoal growth downstream. The importance of floodplain storage of fine sediments should not be underestimated; the absence of overbank deposition on protected lengths of channel maintains higher loads within the channel and may lead to siltation problems.

Most river systems deliver mainly fine sediments to the sea. The River Spey is an exception in that it is a major gravel bed river which actively supplies coarse material to the accreting coastal beach ridges to the south of the Moray Firth. Other than this example, cliff erosion and spreads of marine sediments are the most significant sources of contemporary coarse beach material. The latter source is a legacy of the Pleistocene glaciations and is, therefore, finite.

On the coast, waves and tides generate currents which can transport sediments from source areas to temporary or permanent stores (**sinks**). The transport processes are similar to those operating in rivers and streams, although the sediment pathways are quite different and very complicated. Suspended sediments may be carried many thousands of miles by currents circulating around Great Britain (Figure 2.9). Indeed, cohesive materials supplied by the erosion of the boulder clay cliffs of Holderness is believed to be carried across the North Sea to the coast of continental Europe (Pethick, 1992; see Figure 8.1). Deposition of suspended sediments generally occurs in calm waters of sheltered bays and in estuaries where **mudflats** and **saltmarshes** can develop (see Chapter 8).

Non–cohesive materials are generally transported on the sea bed, either in a **longshore** or **cross–shore** direction depending on the pattern of currents and dominant wave direction. In general, non–cohesive sediments are not carried as far as some suspended sediments and, occasionally, may be confined to a single bay. Transport of coarse sediment supplies the enormous variety of sand and shingle structures that occur around the coast, including beaches, shingle ridges and sand dunes. However, it is important to stress that most of these forms are temporary sediment stores which can release material to the nearshore environment during storm conditions. The continued survival of

17

these landforms is, therefore, dependent on balance between supply and loss; disruption to the supply by, for example, groynes or breakwaters can lead to the gradual degradation of the landform.

The coastline itself is in a delicate balance between the wave energy, tidal regime, resistance of the materials and the supply of sediment. Where there is a small tidal range of less than 3m, as on the east coast of Norfolk, the wave action is concentrated over a narrow range (Figure 2.10).

Figure 2.8 The limits of ice sheet advance in Great Britain

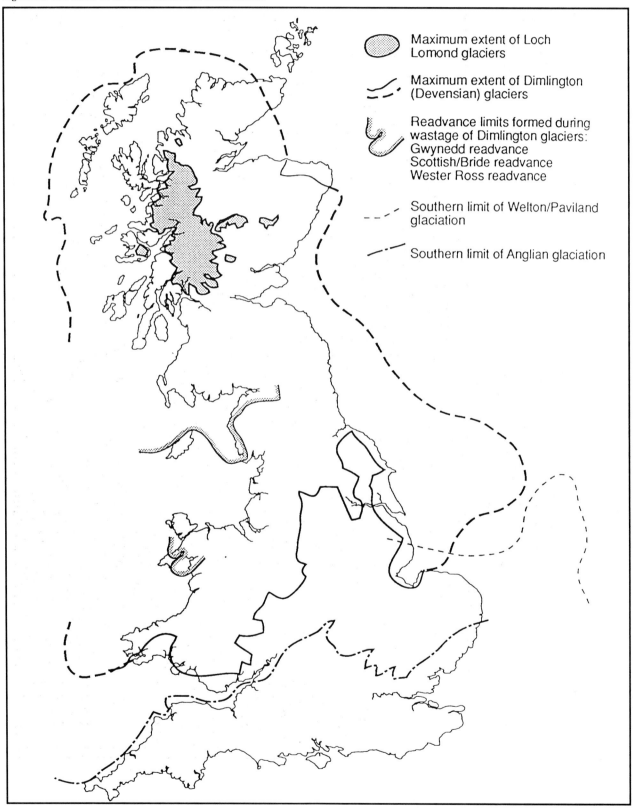

Maximum extent of Loch Lomond glaciers

Maximum extent of Dimlington (Devensian) glaciers

Readvance limits formed during wastage of Dimlington glaciers:
Gwynedd readvance
Scottish/Bride readvance
Wester Ross readvance

Southern limit of Welton/Paviland glaciation

Southern limit of Anglian glaciation

Figure 2.9 The general pattern of water movement around the British Isles (after NERC, 1991)

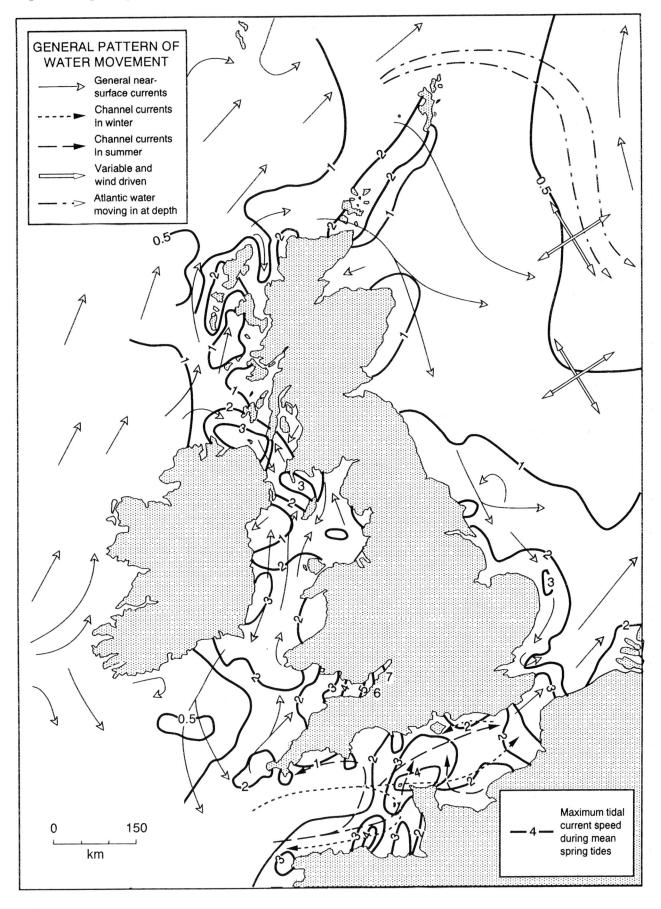

GENERAL PATTERN OF
WATER MOVEMENT

General near-
surface currents

Channel currents
in winter

Channel currents
in summer

Variable and
wind driven

Atlantic water
moving in at depth

0   150

km

Maximum tidal
current speed
during mean
spring tides

Figure 2.10 The tidal range around the British Isles (after NERC, 1991).

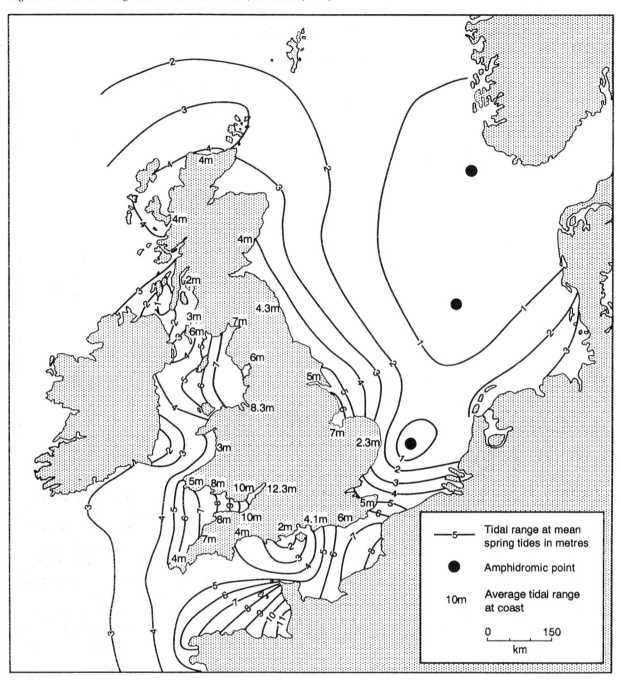

This makes the waves more effective in shaping the shoreline and features such as sandy beaches and spits dominate (Figure 2.11). In areas with larger tidal ranges, as in Lincolnshire, tide-related landscapes dominate with features such as saltmarshes and mudflats.

Processes and Mechanisms: Flooding

Floods occur when water levels rise to overflow land not normally submerged; this can occur in a variety of settings from "dry valleys" to river floodplains and coastal lowlands. Although floods can result from a variety of factors (Figure 2.12), the most common causes are rainfall and snowmelt. Indeed, it is self evident that climatic conditions are the dominant control on flooding, influencing the magnitude and frequency of events and their timing. Climate is of such fundamental importance that it is necessary to briefly outline the broad patterns that determine the flood character of Great Britain.

20

Figure 2.11 The relationship between tidal ranges and landforms around the coast of Great Britain (after Pethick, 1984)

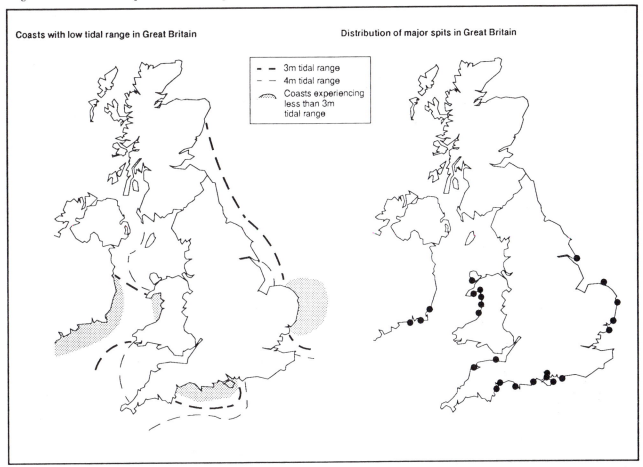

Coasts with low tidal range in Great Britain

Distribution of major spits in Great Britain

- — 3m tidal range
- — 4m tidal range
Coasts experiencing less than 3m tidal range

There are important regional and local differences in climate and flood character which reflect a range of factors, including latitude, proximity to the sea, altitude, aspect and exposure. In general, the west is wetter than the east, with average annual rainfall ranging from over 2500mm in highland and upland areas to less than 750mm in the lowlands of England (Figure 2.13). The annual number of rain days increases from around 165 days in south east England to over 230 days in the Scottish Highlands. Altitude acts to intensify these trends, increasing the amounts of rain associated with frontal depressions and unstable airstreams. For example, the mean annual rainfall at sea level on the west coast is around 1140mm, rising to over 3800mm in the mountains of Scotland, the Lake District and Wales. Even low hills such as the Chilterns and South Downs cause a rise in rainfall, receiving around 120–130mm more in a year than the surrounding lowlands.

Snow fall and the subsequent thaw when temperatures rise again can be important factors in the occurrence of floods. For example, in early 1947 snow fell in some part of Britain on every day from January 22 to 17 March, indirectly causing the great floods of that year when the snow melted (see Chapter 7). Snow fall patterns are also strongly linked to altitude. Close to sea level on the west coast there are on average five days a year with snow falling; the frequency increases by about one day per 15m of elevation between 60–300m and more rapidly on higher ground, giving average values of around 90 days per year at 900m. Average days with snow lying on the ground range from 5 days per year in south west England to over 60 days in the Grampian mountains (Figure 2.14).

The occurrence of exceptionally heavy rainfall has an important bearing on the distribution of flood events. It is interesting to note, therefore, that a daily rainfall of 100mm of more has been surpassed in many parts of the country (Figure 2.15). Indeed, although the heaviest falls have occurred in the north and west, there have been events of a similar magnitude in areas of low annual rainfall, especially near the east coast and in the south west where some of the heaviest daily falls recorded in Britain have occurred. These include:

- Martintown, Dorset; 297.4mm on 18 July 1955.

Figure 2.12 The causes of floods (after Ward, 1978)

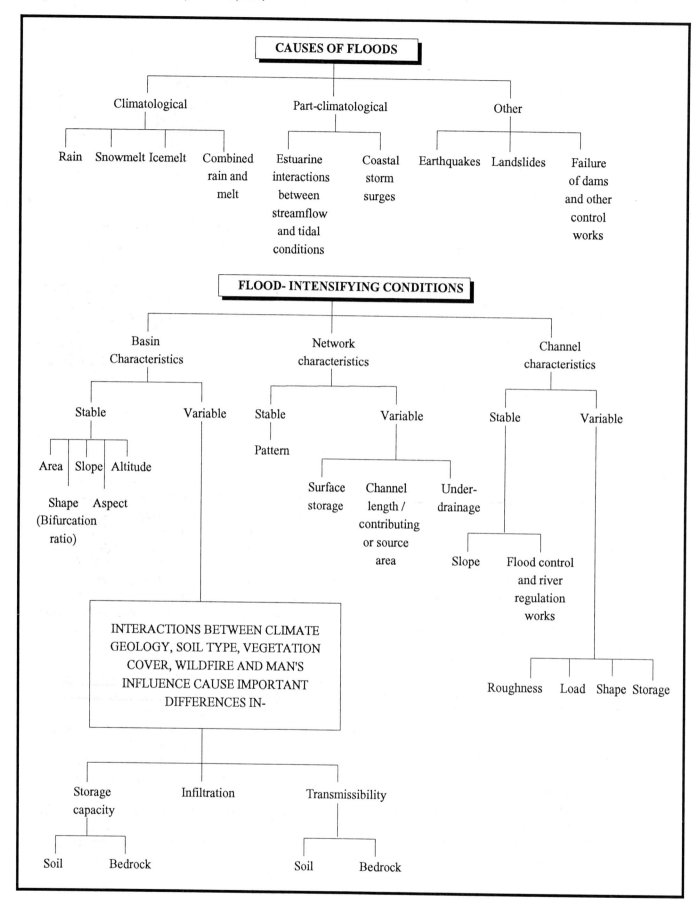

Figure 2.13 The average annual precipitation around Great Britain

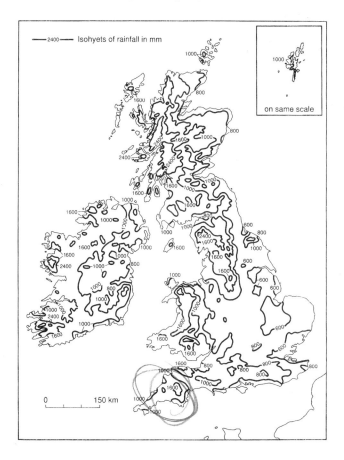

Figure 2.14 The 5-year water equivalent of lying snow (mm) reduced to sea level.

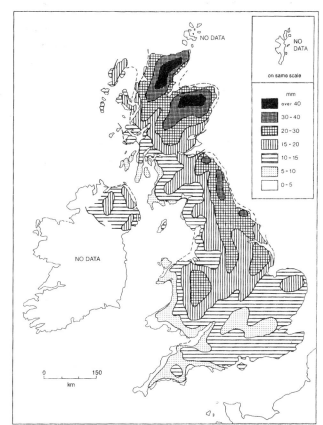

● Bruton, Somerset; 242.8mm on 28 June 1917.

● Cannington, Somerset; 238.8mm on 18 August 1924.

Figure 2.16 shows the return period of a daily fall of 100mm across Great Britain and indicates that these events can be expected, on average, once in every five years or less in many upland areas. For the rest of the country, daily falls of 100mm can have a return period in excess of 100 years. In many areas a rainfall of 100mm or more may be 2 to 4 times the next largest event and can, therefore, have a theoretical return period of thousands of years (as demonstrated by the extreme event at Bath on 10 July 1968; Figure 2.17). However, it is important to note that these return periods refer to the likelihood of such an event falling at a particular rainfall station. Between 1863–1960 around 55 independent falls of over 150mm were recorded; at a national level, an event of this magnitude could be expected somewhere in the country every other year.

These daily totals disguise the intensity of the extreme short duration storms that have been reported (Figure 2.18). For example, during the Hampstead storm of 14 August 1975 169mm fell in 2.5 hours; similar values were thought to have occurred in West Yorkshire on 19 May 1989 (193mm in two hours). Such intense storms generally affect only small areas, although the resultant flash floods can have a devastating impact in small, steep catchments, (e.g. the Lynmouth floods of 15 August 1952, see Chapter 5).

Although rainfall or snowmelt are generally an essential pre-requisite (except for dam failures or on the coast), the actual cause of a flood is the inability of a river or stream to carry all the water flowing through its channel network at a particular location. Amongst the most important factors include the following **regime characteristics**:

● the **catchment area** which affects the total volume of streamflow generated by a catchment-wide event, i.e. the larger the catchment the greater the potential rainfall input;

23

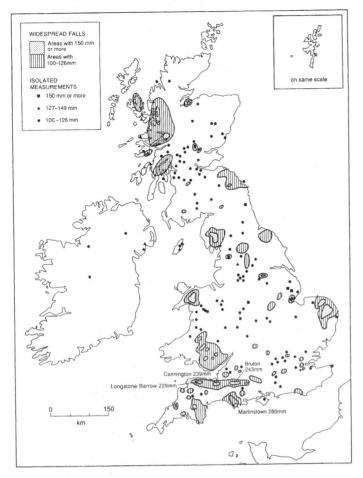

Figure 2.15 The distribution of the largest daily rainfalls recorded in Great Britain (after Rodda, 1970)

- **slope characteristics** which influences the amount of runoff produced by an event, i.e. slope angle, bedrock geology, soil type, land use and vegetation cover;

- **network characteristics** which influence the speed at which water is transmitted through the channel system. In general, the **time of rise** of floodwaters after a rainstorm or melt event will be determined by a range of factors related to the nature of the drainage network itself. **Dendritic networks** tend to produce a marked concentration of flow in the lower catchment as floodwaters are delivered down the major tributaries at a similar speed; **trellised networks** tend to produce a more muted response. The number and size of lakes, reservoirs or other storage areas can be a significant factor in reducing the size of a flood. By contrast, artificial drainage, such as field drains, helps speed up the movement of water towards the channel network.

- **channel characteristics** which influence the ability of the channel to carry a flood flow. Channel capacity is not constant; deposition of eroded sediments can significantly reduce the channel depth and cross section. Entrapment of debris behind structures may cause "backing–up" behind these temporary dams, and lead to overtopping of the banks.

- **antecedent conditions** which determine the amount of the catchment that is saturated prior to a rainstorm or snow melt event and, hence, the amount of runoff that is generated. The river or stream level prior to an event is also critical as this will influence whether the channel system can carry the additional runoff.

The hydrogeological characteristics of a river flood can be described with reference to the **flood hydrograph**, the continuous trace of discharge over time (Figure 2.19). Flooding does not begin immediately at the onset of rain, as the initial increase in discharge is contained within the channel system. The rate of water level rise, the magnitude of the peak flow and duration of flooding are central in defining the nature of a flood event and its impact; these attributes can vary significantly from "flashy" streams which can have a high peak and short duration and "sluggish" streams which may have a relatively low peak and long duration.

In other types of flooding, climate is only partly or indirectly responsible. In many estuaries flooding is often caused by the ponding back of high river discharges by rising tides. On low–lying coasts flooding may result from a combination of high tides and storm–surges generated by high winds and low atmospheric pressure (see Chapter 9).

The Controls of Erosion, Deposition and Flooding Events

The general perception tends to be that erosion, deposition and flooding events are associated with severe storms. Indeed, there is a long and well documented history of **great storm events** around Britain, as indicated by the succession of flood storm disasters that have been recorded throughout previous centuries (Lamb, 1991). These great storms can also have profound permanent or temporary effects on some landscapes:

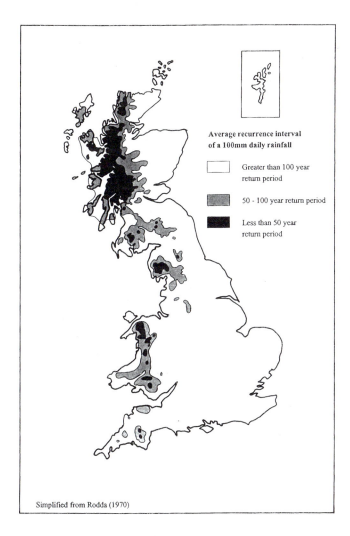

Figure 2.16 The return period of a 100mm daily rainfall in Great Britain (after Rodda, 1970).

Average recurrence interval of a 100mm daily rainfall

☐ Greater than 100 year return period

▨ 50 - 100 year return period

■ Less than 50 year return period

Simplified from Rodda (1970)

- the formation and shifting of sand dune systems;
- the scouring by winds of dry soil or sand and the creation of extensive dust storms;
- the formation of bars and spits across harbour mouths;
- the migration of stream and river channels;
- the initiation of major coastal landslides.

During the January 1953 storms on the east coast, unprotected soft rock cliffs were cut back by up to 30m, Blakeney Point, Norfolk was moved 35m inland and the shingle bank at Albeburgh was flattened (Grove, 1953). There are records of blown sand burying churches on the west and north coast of Cornwall during the 19th Century; at St Enodock a church was buried during a fortnight of storms (Lamb, 1991). Hurricane Charley in August 1986 resulted in one of the most widespread floods recorded this century in northern England. It was accompanied by channel incision and deposition on many of the region's rivers, and floodplain aggradation for long distances downstream (Newson and Macklin, 1990).

It is important to recognise, however, that the processes can also be the product of an unfavourable combination of land management practice and relatively minor rainfall events (e.g. for hillslope erosion, see Chapter 3). Erosion and deposition can be a slow progressive process associated with unexceptional seasonal conditions; here, the rate of the process is as much a factor of topography, geology, soil type and land use as climatic events. Significant events may also occur in response to the operation of "normal" physical processes which slowly change the character of a landform until it reaches a critical level at which, for example, minor rainfalls may trigger major events (e.g. sedimentation in rivers can lead to a progressive reduction in channel capacity and, hence, an increase in flood risk, see Chapter 7). In other instances, accelerated erosion or deposition may be in direct response to the disruption of sediment transport processes by, for example, constructing a dam across a river or groynes along a coastline.

The pattern of erosion, deposition and flooding is often superimposed on a general trend related to the long-term development of a landform, river system or coast. For example, a river channel may be **aggrading**, building up the channel and floodplain, or **degrading** with the incision of the channel and the production of terraces. At present a number of British rivers appear to be incising into their floodplains, with the net removal of coarse and fine material from sediment sinks. This trend is believed to be related to a combination of cessation of metal mining in upland areas which supplied considerable volumes of sediment to rivers and allowed floodplain aggradation, and the result of changing river regimes following a climatic shift to less stormy conditions since the "Little Ice Age" period of colder, wetter conditions between the 16th and 18th centuries.

The climatic deterioration of the "Little Ice Age" emphasises that environmental conditions are not constant. Indeed, climate changes have occurred over a wide range of time-scales, including minor fluctuations, such as the period of warming that

Figure 2.17 Extremes of rainfall at Bath (1911 – 1960) showing the return period of the fall of 10 July 1968 (after Rodda, 1970).

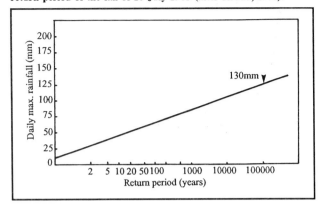

took place in the early 20th century, changes like the Little Ice Age with duration of hundreds of years and the cycles of glacials and interglacials which occurred during the Pleistocene (Figure 2.20). These climate changes are likely to have caused considerable changes in the pattern and frequency of erosion, deposition and flooding events. For example, the Little Ice Age is widely recognised to have been marked by a greater frequency of flooding, wind erosion and coastal landsliding (e.g. Grove, 1972; Lamb 1991). The prospect of global warming and sea level rise suggests that the pattern of events over the next 50 years could differ significantly from that experienced in living memory.

Human activity can have a profound effect on hillslopes, rivers and the coast and, hence, may be instrumental in modifying the nature and frequency of erosion, deposition and flood events. Although human impact on the environment has been of major importance for at least 2000 years, the scale and intensity of disturbance has increased dramatically since the Industrial Revolution. Indeed, rapid urbanisation, mineral extraction,

infrastructural development, water supply, coastal development and dramatic changes in rural land use have all had a significant influence on the erosion, deposition and flood character of many parts of the country, as will be outlined throughout the report.

In summary, the causes of erosion, deposition and flooding are clearly complex and can involve:

(i) **controlling factors** which determine the erosion, deposition and flooding character of an area. These may include sea level, climate, topography, geology and land use;

(ii) **preparatory factors** which make an area susceptible to significant erosion, deposition and flooding events without actually initiating it, e.g. inappropriate land management practices or engineering works which disrupt sediment transport processes;

(iii) **triggering factors** which initiate an event. These are generally related to climate, but may also include engineering operations on coastal slopes (see Chapter 10).

This assessment of the factors influencing erosion, deposition and flooding processes emphasises a number of important points:

- events are often the result of the variable interaction of a range of factors;

- individual factors may have both short-term and long-term influences on the pattern of events. Thus, the effects of development may be immediate (e.g. excavations on coastal cliffs, see Chapter 10) or may take many years to become apparent (e.g. the cumulative effects of floodplain development on reducing flood storage capacity, see Chapter 7);

- long-term environmental changes are important in controlling the erosion, deposition and flooding character of an area (see Chapter 12).

The Geographical Context

Erosion, deposition and flooding should not be seen as merely the **effects** of, for example, heavy rainfall or high seas; they can be **causes** in their own right. River and coastal erosion may oversteepen slopes and lead to instability, erosion

Figure 2.18 Some extreme historical rainfalls recorded in Great Britain (after Acreman, 1989).

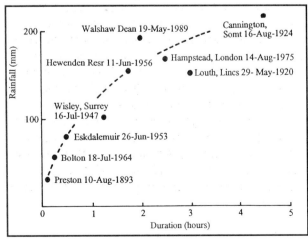

Figure 2.19 The flood hydrograph (after Ward, 1978).

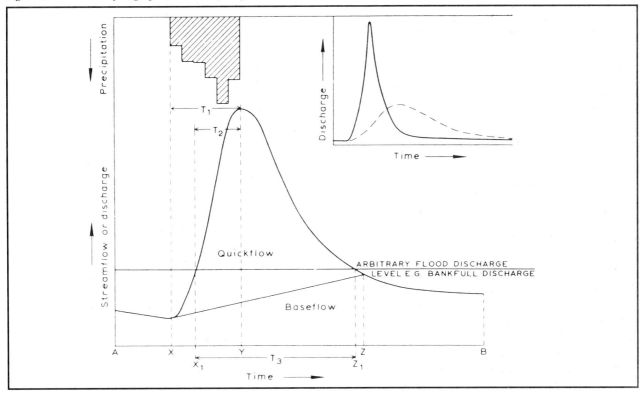

of dunes, beaches, saltmarshes or mudflats can reduce the effectiveness of "natural" and engineered coastal defences and may result in more frequent flooding. In turn, the large volumes of fast flowing water in many floods can cause extensive erosion and, when velocities drop, considerable deposition elsewhere. Channel deposition can increase flood risk.

In considering erosion, deposition and flooding it is important to be aware of how the operation of a process at a particular location can have an effect elsewhere. This can be illustrated by two simple examples;

(i)   **hillslope erosion** can lead to river channel sedimentation downstream, resulting in flood problems or navigation difficulties (Figure 2.21);

(ii) **coastal landslides** may result in lobes of debris and the foreshore, disrupting the transport of sediment to beaches and increasing the likelihood of flooding in lowlying areas or erosion on adjacent cliffs;

The causes and effects of erosion, deposition and flooding are complex and should be viewed as expressed of the operation of large physical systems and not as isolated processes. Indeed, whilst the effects are readily apparent and well appreciated at a local level, they are not often regarded as the product of broader controls that influence the behaviour of river catchments or coastal cells. For example, to understand the flood character of a river, something must be known of the climatic, geological, topographic and land use controls on the supply of water and sediments from the surrounding hillslopes. On the coast, the development of beaches or shingle banks need to be seen as the product of sediment transport within dynamic coastal systems.

To earth scientists and engineers river catchments and coastal systems are a logical spatial framework for considering the various factors that influence the frequency and distribution of erosion, deposition and flood processes (Figure 2.22). For planners and other environmental managers the concepts highlight the need to consider these problems in a wider context than the area covered by a single authority. Indeed, the principle that decision–making should be based on an awareness of both the local conditions (**the site**) and the character of the broader physical system (**the situation**) is of major importance in ensuring that adequate precautions are taken to mitigate against the effects of erosion, deposition and flooding and that development and land use itself does not intensify the level of risk to properties elsewhere in the system.

27

Figure 2.20 Climate variations in Great Britain since 1690 (after Goudie, 1983).

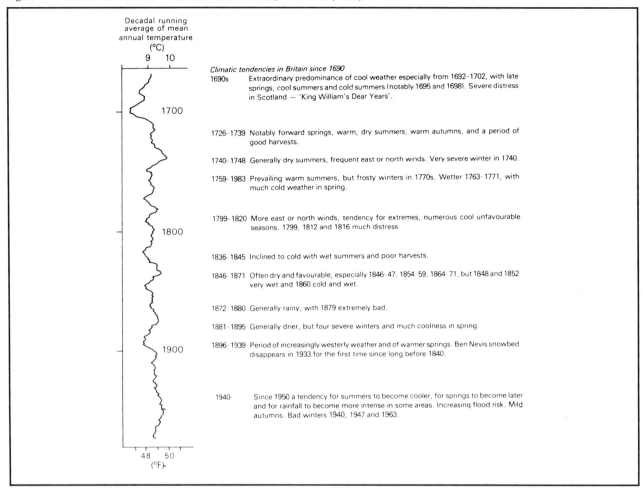

Decadal running
average of mean
annual temperature
(°C)

Climatic tendencies in Britain since 1690

1690s   Extraordinary predominance of cool weather especially from 1692–1702, with late springs, cool summers and cold summers (notably 1695 and 1698). Severe distress in Scotland — 'King William's Dear Years'.

1726–1739   Notably forward springs, warm, dry summers, warm autumns, and a period of good harvests.

1740–1748   Generally dry summers, frequent east or north winds. Very severe winter in 1740.

1759–1983   Prevailing warm summers, but frosty winters in 1770s. Wetter 1763–1771, with much cold weather in spring.

1799–1820   More east or north winds, tendency for extremes, numerous cool unfavourable seasons. 1799, 1812 and 1816 much distress.

1836–1845   Inclined to cold with wet summers and poor harvests.

1846–1871   Often dry and favourable, especially 1846–47, 1854–59, 1864–71, but 1848 and 1852 very wet and 1860 cold and wet.

1872–1880   Generally rainy, with 1879 extremely bad.

1881–1895   Generally drier, but four severe winters and much coolness in spring.

1896–1939   Period of increasingly westerly weather and of warmer springs. Ben Nevis snowbed disappears in 1933 for the first time since long before 1840.

1940–   Since 1950 a tendency for summers to become cooler, for springs to become later and for rainfall to become more intense in some areas. Increasing flood risk. Mild autumns. Bad winters 1940, 1947 and 1963.

(°F)

In addition to providing a framework for understanding the flood behaviour of rivers and streams, the **catchment concept** can help explain the way water and sediments are transported from **supply areas** towards the **coastal zone**. Sediment supply occurs in areas of **hillslope erosion** and where **river channel migration** cuts through areas of stored sediments resulting from past phases of erosion and deposition under different climatic conditions or from an extreme flood in the recent past (e.g. spreads of glacial deposits or floodplain alluvium). The supply of sediment is generally intermittent, with rare floods amongst the most effective events in delivering sediment from these stores into the river channel network. Once in the channel, the sediment size is important in determining how far it is carried before being temporarily stored in features such as point bars or as spreads on the river bed. In short rivers the suspended load may reach the estuary in a single flood, but coarser sediments may become incorporated in the floodplain.

Similar considerations apply on the coast which can be viewed as a series of interlinked physical systems, comprising both offshore and onshore elements. Sediment (clay, silt, sand, gravel etc.), is moved around the coast by waves and currents in a series of linked systems. Simple systems comprise an arrangement of sediment source areas (e.g. eroding cliffs, the sea bed), areas where sediment is moved by coastal processes, and sediment sinks (e.g. beaches, estuaries or offshore sinks). Along a particular stretch of coast these may be a series of such systems, often operating at different scales (Figure 2.23).

In contrast to river catchments, coastal systems have no obvious boundaries. Suspended sediments, for example, may be carried thousands of miles around the coast. Although headlands can be identified which appear to mark the limits of coarse sediment transport they are not permanent boundaries as material may be moved around these sediment divides in severe storm conditions. Despite these difficulties, a series of littoral cells have been identified around the coast of England and Wales (Figure A.1; Appendix A; HR Wallingford, 1993). These cells are based on the movement of coarse sediments and are pragmatic

Figure 2.21 The knock-on effects within a catchment (after Sear and Newson, 1992).

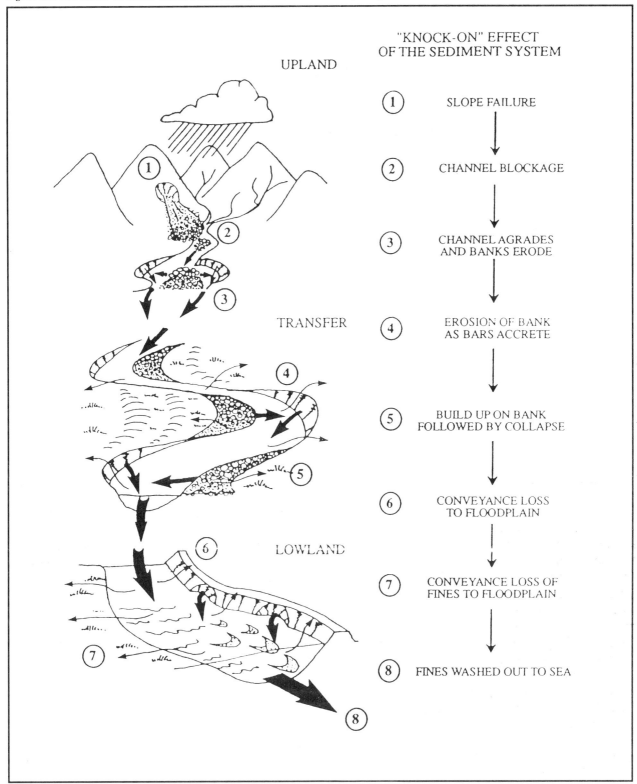

"KNOCK-ON" EFFECT
OF THE SEDIMENT SYSTEM

UPLAND

1. SLOPE FAILURE

2. CHANNEL BLOCKAGE

3. CHANNEL AGRADES AND BANKS ERODE

TRANSFER

4. EROSION OF BANK AS BARS ACCRETE

5. BUILD UP ON BANK FOLLOWED BY COLLAPSE

6. CONVEYANCE LOSS TO FLOODPLAIN

LOWLAND

7. CONVEYANCE LOSS OF FINES TO FLOODPLAIN

8. FINES WASHED OUT TO SEA

tool for coastal defence engineers, identifying lengths of coastline which should be managed in a way that takes account of the interdependence of processes within the cell.

Vulnerable Systems

The response to hillslopes, rivers or coastal systems to the various causal factors can vary considerably. For example, a storm passing over two catchments may result in very different patterns of erosion, deposition and flooding. This **complexity of**

29

Figure 2.22 Catchments and coastal systems.

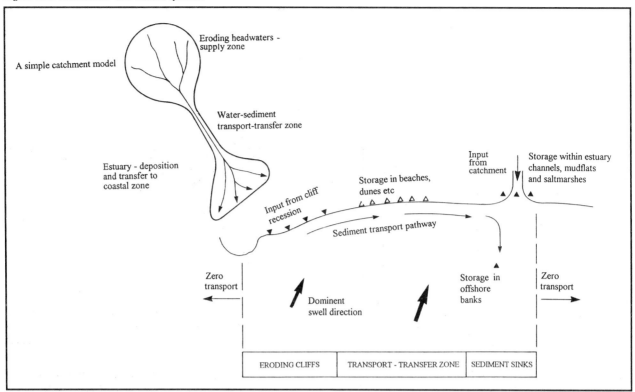

response is a measure of the sensitivity of landform elements within a system which can range from:

(i)   fast responding systems which are very sensitive to disturbing events. This type of system can be morphologically complex because the landforms are subject to rapid change, such as some alluvial river channels and sand dunes;

(ii)   slowly responding insensitive systems, such as hard rock channels or cliffs.

The main controls of sensitivity include material strength, morphological resistance (e.g. relative relief, altitude, slope angle) and the inter–linkages between landforms within a system. A stable coast, for example, is one in which the controlling resistances are sufficient to prevent most storm events from having any significant effects. Thus, although wind and wave energy is greatest on the north and west coasts this does not generally correspond with rapid erosion and sediment transport.

The differing sensitivities to disturbing events gives rise to considerable variation in the erosion, deposition and flood character across the country. The pattern of land use and development is superimposed on this natural variability, with vulnerable areas where human activity is at risk, coinciding with the more sensitive parts of the landscape.

When the pattern of historical records within the **historical records database** (see Chapter 1) is considered in the context of catchments and coastal systems, it is possible to identify those systems where damaging events have most frequently been recorded. These include (Tables A.1 and A.2, Appendix A):

(i)   Catchments

●   the Thames (141 events)
●   the Yorkshire Ouse (91 events)
●   the Clyde (88 events)
●   the Severn (81 events)
●   the Trent (78 events)
●   the Taff (72 events)
●   the Exe (66 events)
●   the Rhymney/Ebbw catchments (59 events)
●   the Usk (51 events)

(ii)   Coastal Systems

●   littoral cell 4; the Thames to Selsey Bill (130 events)
●   littoral cell 3; the Wash to the Thames (85 events)

Figure 2.23 Coastal systems: principal sediment pathways on the south coast of England (after Bray et al, 1992).

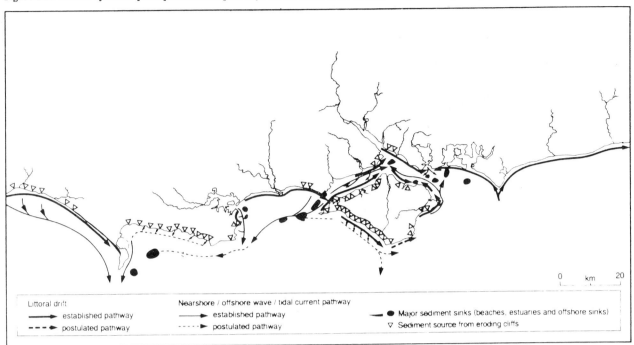

- littoral cell 6; Portland to Land's End (57 events)
- littoral cell 5; Selsey Bill to Portland (35 events)

These catchments and littoral cells clearly may be considered to be prone to erosion, deposition and flood events. However, the reverse is not true; that small numbers of reported events implies a less vulnerable setting. This is because the data sources used to compile the historical record are consistently biased towards developed and densely populated areas. For example, the large numbers of events recorded in the Thames catchment should not necessarily imply that it is more prone to flooding than other rivers.

Predicting Events

Two contrasting approaches are generally used to predict the occurrence of erosion, deposition and flood events:

(i) **deterministic** based on a general understanding of the operation of physical processes. For example, erosion of coastal cliffs is generally modelled through **stability analyses** which compare the destabilising and resisting forces acting within a slope to establish a factor of safety (e.g. Bromhead, 1986). The Flood Studies Report (NERC, 1975) presented a number of deterministic methods of estimating flood discharge from specified rainfall and catchment conditions. However, all deterministic approaches suffer, to a greater or lesser degree, from the need to simplify the complexity of the physical environment and make general assumptions.

(ii) **probabilistic**, regarding individual events as a random part of a natural series of events of varying magnitude and frequency, whose distribution can be established from a sequence of records and whose probability of occurrence in a given period can be calculated using standard statistical methods. This approach is most suited to flooding, where the likelihood of a flood of a particular magnitude is generally expressed by the **return period** or **recurrence interval**. Thus, the flood which is expected to be equalled or exceeded **on average** every 100 years, has a return period of 100 years. This event could occur any year, but the probability of its occurrence during 100 years is much greater than during a one year period.

The relationship between probability (expressed as a percentage), return period and the length of period under consideration is shown in Table 2.1. This indicates that a 1000 year event has a 6% chance of occurring during the lifetime of a building (taken here as 60 years).

31

Table 2.1 Percentage probability of the N-Year flood occurring in a particular period (after Ward, 1978).

Number of years in period	N = Average Return Period, T_r, in years							
	5	10	20	50	100	200	500	1000
1	20	10	5	2	1	0.5	0.2	0.1
2	36	19	10	4	2	1	0.4	0.2
5	67	41	23	10	4	2	1	0.5
10	89	65	40	18	10	5	2	1
30	99	95	79	45	26	14	6	3
60	–	98	95	70	31	26	11	6
100	–	99.9	99.4	87	65	39	18	9
300	–	–	–	99.8	95	78	45	26
600	–	–	–	–	99.8	95	70	45
1000	–	–	–	–	–	99.3	87	64

Where no figure is inserted the percentage probability >99.9

Perhaps the single most important limitation on both deterministic and probabilistic approaches is the limited data sets of, for example, rainfall or flow records from which predictions have to be made. Benson (1960) demonstrated that to achieve 95% reliability on the estimate of discharge of a 50 year flood event required 110 years of records; such lengthy data sets are not common in Great Britain, highlighting the problems associated with defining the likelihood of extreme events that will be a recurrent theme throughout this Report. Detailed rainfall records are usually available for 100–150 years, hydrological records from river gauging stations generally for 20–40 years (one of the longest gauging records cited in the Flood Studies Report (NERC, 1975) is for the Thames at Teddington, beginning in 1883). Wave data are much scarcer (HR Wallingford, 1987) with extreme wave heights often extrapolated from relatively short data sets.

The Significance of Erosion, Deposition and Flooding

The report is not merely a review of the contemporary behaviour of these physical systems; it aims to focus attention on those processes which can, in particular circumstances, impose significant constraints to land use and development. That the processes have had notable impacts on the economy is clearly demonstrated by the records of significant events that form the historical database. These records describe the reported impact of a specific process at a particular location; by collating all the records generated by the same sequence of climatic events, or other causal factors, it has been possible to gain a general indication of the cumulative effects of the different processes operating within the same **erosion, deposition and flooding event**. For example, the passing of an intense storm over Calderdale on 19 May 1989 generated (Acreman, 1989):

(i)   flooding of houses in Luddenden to a depth of more than 1m, sweeping away several cars and caravans;

(ii)   erosion around culverted sections of Luddenden Brook, removing overlying gardens, greenhouses and sheds leaving houses close to a gulley 20m wide and 5m deep.

A total of 1550 separate events have been identified within the historical samples. The breakdown of event frequency for each decade since 1700 is shown on Figure 2.24 and in Table A.3 (Appendix A). The pattern reveals an almost exponential increase in frequency of events up to the 1950s after which the number of events has remained fairly constant. The factors influencing this pattern area likely to be highly complex, but are likely to include:

Figure 2.24 The frequency of events of different magnitude, from 1700 to 1993.

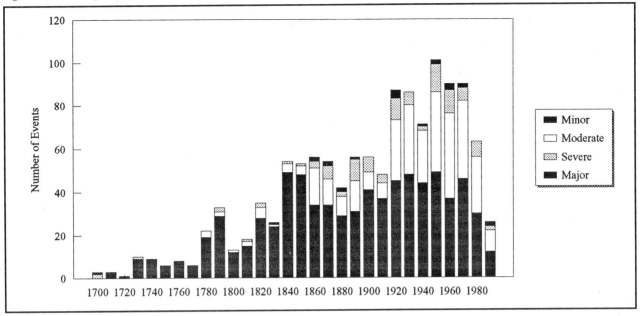

• the rapid spread of development into vulnerable locations throughout the 19th and 20th centuries, resulting in an increase in the risk of a damaging event;

• the significant institutional and structural responses to the major flood disasters of 1947 and 1953 with improved flood warnings, and defence schemes based on a better appreciation of catchment and coastal processes, have resulted in a reduction of risk in protected areas.

Each event comprises a unique collection of records, each with different affected areas, duration and impact characteristics. However, by assessing the cumulative effects of the reported incidents it has been possible to classify each event according to the overall magnitude of the impact. Because of the enormous variety of impacts this has largely been a subjective procedure, although a range of indicative criteria were used to guide the classification process (Table 2.2). The pattern of events of different magnitude reveals the following average frequencies per decade this century:

• **Class 1 events** 40 per decade;
• **Class 2 events** 25 per decade;
• **Class 3 events** 7 per decade;
• **Class 4 events** 1 per decade.

It is important to stress that many of the more serious problems have had some form of treatment to reduce the impact of future events of similar magnitude on the community. For example, since the devastating east coast floods of 1953 there has

been a major programme of sea defence and flood warning improvements which prevented a repetition of the disaster when an even higher storm surge occurred on 11–12 January 1978. It should be noted, however, that **defences cannot eliminate the risk of a damaging event**. For example, flood defences in Perth, Tayside were raised in 1974, to a design–level that had not been exceeded since 1814. However, in January 1993 the defences were overtopped and around 1500 properties affected in and around the city, emphasising that defence schemes only provide protection against events up to a certain size.

Conversely, urban growth and development has been responsible for increasing the risks in certain areas. Most obviously, communities have extended into areas where erosion, slope failure and flooding have always existed. At the same time, urban growth has significantly altered surface characteristics and hydrology. The transformation of agricultural land to a housing estate with the efficient artificial drainage and impermeable concrete or tarmac surfaces profoundly influences the quantity and rate of runoff, increasing both. Events which would have had a minor impact in the past could now affect many sectors of the economy because of the concentration of resources and infrastructure in vulnerable locations. Thus one of the urgent tasks of hazard mitigation is to establish the extent of land vulnerable to those processes and to determine the degree of risk that can be expected, so that adequate precautions can be taken to avoid damage or losses, including loss of life.

Table 2.2 Indicative criteria for the classification of significant events.

MAGNITUDE OF IMPACT	INDICATIVE CRITERIA	
	LOCALISED	WIDESPREAD
Minor Event Class 1	Individual towns and villages suffer flooding with no more than ten houses inundated and four houses destroyed, flooding of minor roads. Localised erosion including damage to bridges.	National flooding of agricultural land and infrastructure; regional traffic disruption. Few communities affected. Less than £0.5M damage in total
Moderate Event Class 2	Intense, localised damage in towns or villages; may involve up to five dead. Considerable local disruption, with financial hardship to a few.	A region's towns and villages suffer flooding with more than 2000 houses flooded. Damage is not intense or widespread within a community. Less than £5M damage in total.
Severe Event Class 3	Considerable localises damage in towns and villages; may involve up to 15 dead. Event may involve lengthy period of inundation and severe damage to individual properties, widespread evacuation and emergency relief.	A region may experience setbacks to industry, financial hardship to a few and financial setbacks to thousands. Cities may be severely inundated in a number of districts; local towns and villages badly affected. Up to 6000 houses flooded, with damage up to £50M.
Major Event Class 4	Almost complete desolation and destruction to a community; may involve up to 30 dead and widespread destruction of property. Evacuation and disaster relief aid required.	Considerable damage to region's cities, towns and villages; may involve over 30 dead. Widespread setbacks and financial hardship to thousands of people. More than 10,000 made temporarily homeless, with damages in excess of £50M.

The review has identified a number of key areas where erosion, deposition and flooding has a significant impact on land use and development:

- slope erosion and mudfloods on hillslopes;
- dust storms on hillslopes;
- flash floods in upland areas;
- bank erosion, sedimentation and channel instability on rivers;
- floods on lowland rivers;
- sedimentation in estuaries;
- floods in low-lying coastal areas;
- erosion of coastal cliffs;
- wind blown sand in coastal dunes.

In the following Chapters 3 to 11 each of these nine key problems are discussed, providing a national overview on their occurrence and significance in terms of:

- the nature of the problems;
- causes;
- the identification of vulnerable areas;
- the effects of development;
- the significance to conservation;
- the significance to land use planning.

The final Chapter attempts to summarise the overall order of costs that can result from these problems and also the benefits that may arise. It examines the way the occurrence of the problems are linked to the behaviour of physical systems and how socio-economic factors can have an important influence on the scale of impacts. By way of conclusion, the report examines the likely effects of global warming and sea level rise with reference to the pattern of significant events over the last 200 years and more.

Chapter 2: References

Acreman M C 1989. Extreme rainfall in Calderdale, 19 May 1989. Weather. 44, 438–445.
Benson M A 1960. Characteristics of frequency curves based on a theoretical 1000 year record. US GS Water Supply Paper 1543–A.
Bray M J, Carter D J and Hooke J M 1992. Coastal sediment transport study. Reports to SCOPAC. Dept. of Geography, Univ. Portsmouth.
Bromhead E N 1986. The Stability of Slopes. Surrey Univ Press.
Dearman W R and Eyles N 1982. An engineering geological map of the soils and rocks of the UK.

Bulletin of the International Association of Engineering Geology, 25, 3–18.

Grove A T 1953. The sea floods on the coasts of Norfolk and Suffolk. Geography. 38, 164–170

Grove J M 1972. The incidence of landslides, avalanches and floods in western Norway during the Little Ice Age. Arctic and Alpine Research. 4. 131–138.

H R Wallingford 1987. Wave data around the coast of England and Wales. A review of instrumentally recorded information. Report SR113.

H R Wallingford 1993. Mapping of littoral cells. Report SR 328.

Jones D K C and Lee E M 1994. Landsliding in Great Britain. HMSO.

Lamb H H 1991. Historic storms of the North Sea, British Isles and Northwest Europe. Cambridge University Press.

Morgan R P C 1980a. Soil erosion. Longman.

Morgan R P C 1980b. Soil erosion and conservation in Britain. Progress in Physical Geography, 4, 24047.

NERC 1975. Flood Studies Report. Institute of Hydrology, Wallingford.

Newson M D and Macklin M G 1990. The geomorphologically–effective flood and vertical instability in river channels: a feedback mechanism in the flood series for gravel bed rivers. In W R White (ed) River Flood Hydraulics. J Wiley and Sons.

Pethick J 1984. An introduction to coastal geomorphology. Arnold Press.

Quine T A and Walling D E. Rates of soil erosion on arable fields in Britain : quantitative data from caesium – 137 measurements. Soil Use and Management 7, 169–176.

Rodda J C 1966. A study of the magnitude, frequency and distribution of intense rainfall in the United Kingdom. British Rainfall 204–215.

Rodda J C 1970. Rainfall excess in the United Kingdom. Transaction of the Institute of British Geographers, 49, 49–59.

Sear D A and Newson M D 1992. Sediment and gravel transportation in rivers including the use of gravel traps. NRA project report. 232/1/T.

Goudie A 1986. The human impact on the natural environment. Blackwell.

Goudie A 1990. The landforms of England and Wales. Blackwell.

Lewin J (ed) 1981. British rivers. George Allen and Unwin.

Morgan R P C 1980. Soil erosion. Longman.

Newson M D 1975. Flooding and flood hazards in the United Kingdom. Oxford University Press.

Pethick J 1984. An introduction to coastal geomorphology. Arnold Press.

Penning–Rowsell E C, Parker D J and Harding D M 1986. Floods and Drainage. George Allen and Unwin.

Perry A H 1981. Environmental hazards in the British Isles. George Allen and Unwin.

Schumm S A 1977. The fluvial system. Wiley Interscience.

Smith K 1992. Environmental hazards : assessing risk and reducing disaster. Routledge.

Ward R C 1978. Floods : A geographical perspective. Macmillan Press.

Chapter 2: Suggested Reading

Carter R W G 1988. Coastal environments. Academic Press.

Cooke R U and Doornkamp J C 1990. Geomorphology in environmental management. Oxford University Press.

Goudie A 1983. Environmental change. Oxford University Press.

3 Hillslopes: Slope erosion and mudfloods

The Nature of the Problems

Surface water flow (**runoff**) occurs when the rainfall intensity during a storm exceeds the rate of infiltration into the soil or when the soil is saturated. Surface flow occurs as a continuous sheet of runoff (**overland flow**), in small channels (**rills**), or larger, more permanent features (**gullies**, Figure 3.1). Towards the foot of a slope the flow may concentrate in a permanent watercourse or an ephemeral channel such as a "**dry valley**". In some circumstances, the development of networks of pipes in the upper layers of the soil can contribute to runoff. Hillslope runoff can **transport** soil particles, provided they have been **detached** from the soil mass by, for example, rainsplash. Considerable quantities of eroded material can be carried in surface runoff and may result in "**mudfloods**" or "**mudflows**". Deposition occurs in fans or spreads of debris, usually at the base of the slope, when the power of the water is no longer sufficient to carry the soil particles.

Problems can arise in both upland and lowland Britain. Whilst it would be wrong to be complacent about the significance of hillslope erosion to land use planning, it should be recognised that many of the following examples are primarily **agricultural, forestry and conservation issues** and not planning matters. However, soil erosion can have a number of "knock–on" effects outside the eroding hillslopes or fields and, hence, can indirectly influence planning objectives in an area. Although the planning system would not generally be an effective mechanism for achieving reductions in soil erosion, it is important that planners are aware of the linkages between land use and hillslope processes, catchment management and river corridor problems such as bank instability and flooding.

In **upland areas** the growing popularity of rambling and long distance walking has led to increased fears of erosion along existing rights of way. For example, a recent Countryside Commission report identified 30 stretches of the Pennine Way that were severely eroded and in need of major restoration at a cost of over £600,000. In 1990, 120 tonnes of sandstone had to be lifted by helicopter to the summit of Ingleborough, in the Yorkshire Dales National Park, to rebuild eroded footpaths as part of an £800,000 restoration programme.

Afforestation operations in upland areas can cause increased erosion and lead to more sediment in upland streams. Particular problems can result from ploughing, harvesting and roadmaking operations which expose the soil surface and tend to increase the speed of runoff. Table 3.1 summarises the results of a number of catchment studies, highlighting observed increased stream sediment loads of up to 1600% in forested areas compared with unafforested areas. The issue is a controversial one, with considerable difference of opinion over its significance between the forestry industry (e.g. Moffatt, 1988) and nature conservancy agencies (e.g. Soutar, 1989). Of particular concern is the possible impact of increased sediment loads on salmon and trout rivers. For example, Drakeford (1979, 1982) correlated the reduction in catches of salmonoid fish in the River Fleet with the expansion of forestry in the catchment, noting a 90% decline in sea trout catches between 1960 and 1978. It should be noted, however, that forestry practice has improved over the last decade with the Forestry Commission (1988) guidelines bringing together previous advice aimed at reducing the impact of afforestation on fresh waters.

Erosion of upland peat is also seen as a serious threat to many valued landscapes, especially in the Pennines. The subject has generated considerable academic interest with numerous studies into the nature and causes of peat erosion (e.g. Bower 1960a, b; Bower 1962; Tallis 1964, 1965; Crisp 1966; Phillips et al, 1981). In recent years attention

Figure 3.1 The main factors in soil erosion by water (after Cooke and Doornkamp, 1990).

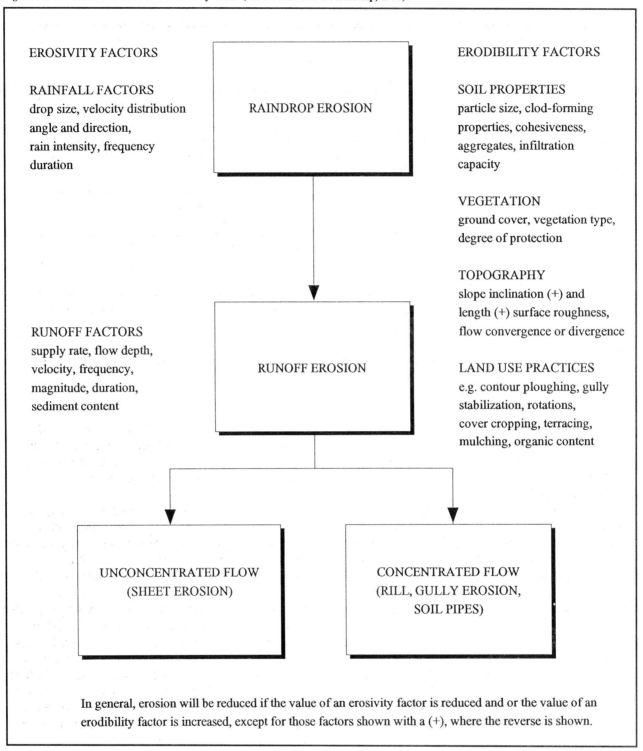

EROSIVITY FACTORS

RAINFALL FACTORS
drop size, velocity distribution
angle and direction,
rain intensity, frequency
duration

RAINDROP EROSION

ERODIBILITY FACTORS

SOIL PROPERTIES
particle size, clod-forming
properties, cohesiveness,
aggregates, infiltration
capacity

VEGETATION
ground cover, vegetation type,
degree of protection

TOPOGRAPHY
slope inclination (+) and
length (+) surface roughness,
flow convergence or divergence

RUNOFF FACTORS
supply rate, flow depth,
velocity, frequency,
magnitude, duration,
sediment content

RUNOFF EROSION

LAND USE PRACTICES
e.g. contour ploughing, gully
stabilization, rotations,
cover cropping, terracing,
mulching, organic content

UNCONCENTRATED FLOW
(SHEET EROSION)

CONCENTRATED FLOW
(RILL, GULLY EROSION,
SOIL PIPES)

In general, erosion will be reduced if the value of an erosivity factor is reduced and or the value of an
erodibility factor is increased, except for those factors shown with a (+), where the reverse is shown.

has focused on the implications for landscape and nature conservation interests, together with the potential effects on reservoir sedimentation (e.g. Labadz et al, 1991).

Although hillslope erosion in upland areas may lead to land degradation and loss of grazing land or amenity, probably the most pressing environmental issues are related to the effect on upland water management:

- the loss of water storage capacity, as soil and peat removed from moorlands is deposited in reservoirs;

- decline in water quality due to the high sediment yields from eroding hillslopes;

- sedimentation in watercourses, leading to a reduction in capacity and, hence, increased flood risk and navigation problems (see

Table 3.1 Sediment yields associated with forestry in upland Britain (after Soutar, 1989).

Catchment	Sediment Yield as Suspended Sediments (kg ha^{-1} yr^{-1})	Afforested or Unafforested	Comment	Authors
Llanbrynmair catchment Wales	37 90	Unafforested After ploughing	Suspended sediments increased to 246% of former levels after ploughing.	Francis & Taylor (1989)
Hore, Wales	244 571	Mature afforestation After felling	Suspended sediments increased to 234% of former levels after felling.	Leeks & Roberts (1987)
Coalburn	30 240 120	Unafforested Average yield over first five years after ploughing Longer–term yield after ploughing	Suspended sediments increased to 800% of former levels in first five years after ploughing. This later fell to 400% of former pre–afforestation levels.	Robinson & Blyth (1982)
Balquhidder Monachyle Balquhidder Kirkton	380 1310	Unafforested moorland Afforested	Total sediments in afforested catchment were 349% of those in unafforested catchment.	Ferguson & Stott (1987)

Chapters 7 and 8).

It was shown recently that loss of capacity in a sample of southern Pennine reservoirs varied between 4% and 75% in around 100 years since construction (Butcher et al, 1992). Elsewhere, records provided by Northumbria Water indicate a variable pattern with loss of capacity generally less than 15% over 100 years (Table 3.2). In recent years concerns have been expressed over the impact of hillslope erosion on the discolouration of water supplies, increasing the cost of water quality treatment (Edwards et al, 1987).

It has been estimated that between 37% and 45% of **arable land** in England and Wales is at risk from soil erosion, (Morgan, 1985; Evans and Cook, 1986). In any one year between 5% and 15% of arable land may erode (Evans, 1988). Typical rates of annual soil loss are reported to be around 1–5 tonnes/ha (Boardman, 1990), although single event losses of 20–40 tonnes/ha may occur with a return period of 1–5 years (Morgan et al, 1987; Evans, 1988); extreme events generating in excess of 100 tonnes/ha are not uncommon (Evans, 1988; Boardman, 1990).

The impact of erosion and deposition in lowland areas can also be divided into **on–farm** and **off–farm** impacts. The relationship between the two types of impact is not straightforward; an event

with serious on–farm impacts may have no off–farm consequences either because the impacts are contained within the farm or because of a lack of vulnerable off–farm sites in the vicinity.

On–farm impacts include immediate damage to crops, loss of fertilizer, crop burial, difficulties in harvesting gullied fields, damage to farm yards and buildings (Boardman, 1990b, 1992b). It also includes long–term loss of yield because of soil thinning, especially loss of organic matter and resultant loss of water and nutrient holding capacity of the soil. Over the long term there will also be textural changes in the soil due to preferential loss of fine particles and incorporation by ploughing of subsoil or bedrock material. Since British soils tend to contain a highly beneficial component of wind–blown silt (loess) in their upper 1m, loss of the upper horizons inevitably means a deterioration in soil quality. However, it is important to bear in mind:

- rates of erosion in Britain are generally low therefore have had little effect on yield over the time scale of decades;

- any possible decline in yield through erosion is masked by application of fertilizers and improvements in strains of arable crops.

Table 3.2 The effects of sedimentation in reservoirs in north-east England (source: Northumbria Water).

Reservoir	Years of Operation	Elevation m.AOD	Loss of Capacity %
Hury	98	262	1.9
Blackton	96	282	11.7
Balderhead	27	333	1.1
Grassholme	77	275	8.9
Selset	32	316	4.7
Cow Green	22	489	1.2
Lockwood	124	190	13.4
Scaling	37	185	1.8
Burnhope	55	400	5.6
Waskerley	113	360	6.2
Tunstall	112	220	8.8
Hisehope	86	340	12.5
Smiddy Shaw	115	340	6.5
Bakethin	10	185	3.2
Kielder	10	185	1.1
Fontburn	90	190	14.5

There has been little research on this topic in Britain, although Frost and Speirs (1984) reported a detailed study of two fields in south east Scotland suggesting that:

"current rates of erosion at this site can be tolerated for at least 200 years before the land capability for agriculture is significantly and permanently reduced by droughtiness or stone content." (Frost and Speirs, 1984).

Information on the **off-farm impacts** of hillslope erosion and runoff events is scarce but may include clearance of soil from roads and ditches; damage to houses and gardens caused by mudfloods; siltation of waterways, lakes and reservoirs; decline in quality of water supplies; transfer of agricultural chemicals especially phosphorus and pesticides attached to soil particles during erosion and mudflood events. In the South Downs, over 30 separate incidents have been reported since 1976 affecting around 200 houses, with the scale of problems ranging from inundation by soil-laden water to damage to outbuildings and gardens (Table 3.3 summaries some of the more severe events; Boardman, 1990c). Estimates of costs associated with the events suggest a minimum

figure of £836,000 over a 15 year period since 1976. The most significant event was that of October 1987 at Rottingdean, Mile Oak and Hangleton, which resulted in householder costs of £406,000 and local authority costs of £353,000 (excluding the emergency services). Robinson and Blackman (1990) noted:

"the majority of the costs were borne by the local authorities and insurance companies, but uninsured financial losses that had to be borne directly by the householders amounted at least £112,500" (Robinson and Blackman, 1990).

All the recorded incidents occurred in the autumn and winter and are the result of runoff from land under, or prepared for, winter cereals. Of interest, there is no evidence of similar events prior to 1976, when the area of winter cereals was much lower.

Table 3.4 highlights a number of runoff generated events in the Isle of Wight where roads and property have been affected. The problems are confined to sandy soils cultivated for winter cereals, oil seed rape and early potatoes. By far the most serious problem occurred 26–28 October

Table 3.3 Soil erosion and mudfloods: Property damage due to runoff from arable fields on the South Downs (1976–1990).

Location	Problem	Action	Source
Highdown Estate, Lewes.	Flooding of houses and roads, Nov–Jan '82–83. Costs: £12,000 to local authorities.	Lewes DC advised they would not win a legal case against farmer. Farmer changed from winter to spring cereals in following year. No further flooding.	Boardman et al, 1983 Stammers and Boardman, 1984.
Breaky Bottom Farmhouse, Rodmell.	Farm flooded 1976,1982 and 1987. Vineyard also damaged in 1987. Costs: 1987: >£80,000.	Legal action against neighbouring farmer. Settled out of court in 1993.	Boardman 1988b.
Shepherds Mead, Worthing.	Flooding of houses and roads: 1976, 1980, 1982, 1983 and 1987. Runoff from fields farmed by tenant of Worthing BC.	After 1987 flooding farmer agreed to change from cereals to grass with rent reduction by Council. No further flooding.	
Mile Oak.	Flooding of houses and roads 1976.	Reconstruction of one dam and building of three others in 1987.	Robinson and Blackman 1990.
Mile Oak.	Flooding of houses and gardens, October 1987. Three dams collapsed. Water entered 16 houses, 18 garages and 31 gardens inundated. Costs: householder >£106,000, local authority £153,000.	Three dams constructed. Site now protected by new Brighton Bypass embankment.	Robinson and Blackman 1990.
Woodingdean.	Flooding of houses, October 1987.	Ditch, embankment and permanent drainage system constructed by Council.	
Ovingdean.	Floodwater and soil into gardens and houses, October 1987.	Ditch dug by Council to prevent further flooding.	
Rottingdean.	Flooding 7 October 1987, 66 houses flooded. Runoff from farm in tenancy of Brighton BC. Costs: householders c£300,000, local authorities £120,000	Emergency protective measures, thereafter one dam built and land use change agreed with Council. Homeowners advised they would not win legal case to recover costs from farmer.	Boardman 1988b, Robinson and Blackman 1990, Robinson et al 1987.
Polegate.	Houses affected by runoff 5–6 times in 1987, some previous occurrences, 'increasing frequency recently'.	No action taken	
Bevendean.	Flooding of house 1983, 5 October 1987 & February 1988. Garden flooded 15 February 1991. 11 October 1993 garden flooded and water through air bricks and under house; three other houses affected.	Storm drains fitted. Two dams and ditches built by Council, manhole installed to take water to main sewer.	
Pyecombe Golf Club.	Flooding of 10th green by runoff from arable fields. 26–69 October 1990. Costs £25,000	Golf Club constructed new green in low risk location.	Boardman 1990a
Sompting: Steepdown valley.	Flooding of 6 houses in Herbert Road and Valley Road in 1980, 1987 and 1990–91. Costs to Adur DC >£40,000 October 1993, flow of water and soil across road at TQ 163075; Titch Hill farm flooded.	One small dam and one soakaway built by farmer (August 1991). Some land use change to set aside for winter 1991–92. Temporary soil dam built by farmer to stop runoff; second dam higher up basin.	Boardman and Evans 1991, 1993

Table 3.4 Off-farm impacts of runoff and erosion: Isle of Wight.

● Mottistone Manor farm, soil on road, winter 1992-93. Soil from badly eroded winter cereal field.
● Hunney Hill, Brighstone, soil on road, winter 1992-93. Upper Lane and Main Road, Brighstone blocked 26-28 October 1990.
● Blackwater Road, Newport; serious problems 1990 and recurrence 1992-93. Birchfield House : river of mud broke down door leaving ten feet of mud in the cellars, fire service spent seven hours pumping and returned next two days (Isle of Wight County Press, 2 November 1990). Road from Blackwater to Rookley closed 26-28 October 1990.
● Arreton Cross sediment trap at junction of Sunken Lane and main road, emptied every two months. Arreton village 26-29 October 1990: avalanche of mud poured through village, abandoned cars on main road, mud removed from Red Lion Cottages (£500 damage) (Isle of Wight County Press, 2 November 1990).
● Presford Farm, 1km east of Shorwell (SZ 470822), flooding of road 1990, ditch system now constructed to take water under road.
● Ventnor Road, Apse Heath sandbags along road March 1993.
● Whitecroft Hospital, Sandy Lane, Newport 3 October 1984: silt onto Road, cost charged to landowner.
● Pan Lane, Niton blocked 26-28 October 1990 with mud spreading into other village roads.
● Sheepwash Farm, Godshill 26-28 October 1990, avalanche of mud from nearby hills, four feet of slurry in downstairs rooms, thousands of pounds of damage (Isle of Wight County Press, 2 November 1990).
● Taverners Inns, Godshill, 26-28 October 1990, slurry came in back and front doors, six inches of mud, damage estimated at £1000 (Isle of Wight County Press, 2 November 1990).

1990 when 78mm of rainfall was recorded, mainly on the night of the 26th. The worst affected areas were the villages of Brighstone and Blackwater which were both cut off. Sixty properties were affected by flooding and the clean-up costs have been estimated at £25,000 for the County Council and in the order of £2,000-£3,000 for homeowners. At the time there was considerable criticism of farmers in the local press and 35 farmers or landowners were served with notices by the County Council under the Highways Act (1980) S.150. Most farmers have since tried to prevent a recurrence of flooding by the use of straw bales and the digging of ditches.

Similar problems have been recorded throughout England and Wales, although particularly susceptible areas include parts of Devon, Kent, Norfolk, Nottinghamshire, Shropshire, Somerset and Suffolk. It is interesting to note that there appears to have been an increase in frequency of reported events in the mid 1980's, a trend that is believed to reflect, at least in part, the changes in land management that occurred around that time (see below, Figure 3.2).

Causes of Hillslope Erosion

The causes of hillslope erosion are complex, but can be discussed in terms of:

● the **soil type**;
● the **vegetation cover**;
● the **rainfall intensity**;
● the **slope gradient**;
● the **land management** practice.

Runoff occurs when the hillslope soils are saturated (i.e. the **soil moisture storage capacity** is exceeded) or if the rainfall intensity is greater than

Figure 3.2 The relative frequency of reported incidents of hillslope erosion events since 1966.

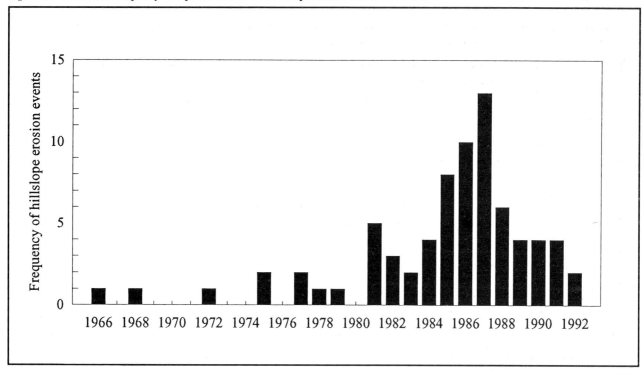

the **infiltration rate** of the surface horizons. In Great Britain, the former process is believed to be the most significant, although it is important to note that compacted and impermeable surfaces may generate runoff volumes close to 100% of the rainfall. Both processes vary according to **soil type**, especially the soil texture (grading) and permeability. Soil loss occurs when there is sufficient runoff over soils prone to the effects of rainsplash. In this context, soil erodibility varies with a variety of physical properties including: soil texture, shear strength and aggregate stability, with silts and fine sands especially prone to erosion (see, for example, Boardman, 1993).

Vegetation cover has an important role in reducing erosion through the interception of raindrops, so that their energy is dissipated by the plants rather than imported to the soil. A ground cover also reduces the velocity of running water and, hence, the potential for erosion. In addition, a good root network can open up a soil, enabling water to penetrate and reducing the generation of runoff.

Soil loss is clearly related to **rainfall** because of its role in detaching particles through the impact of droplets and through its contribution to runoff. Until recently it had been assumed that exceptional soil loss was the produce of intense rainfall events. However, it is now recognised that hillslope erosion events can follow rainstorms of less than 20mm. Indeed, erosion and runoff resulting in property damage may result from relatively modest

rainfall amounts if field surface conditions are suitable (Table 3.5). The 50mm in 24 hours that caused considerable damage in the Bishopsteignton area is estimated to have a return period of less than five years. In the case of the flooding in Ashford-in-the-Water, west Derbyshire, only 6mm of rain fell on the day of flooding.

Although low-intensity/long duration rainfall appears to be responsible for many erosion events, the impact of high intensity storms should not be overlooked. One of the most serious lowland erosion events recorded in Britain occurred in East Sussex in October 1987 as a result of heavy rainfall (63mm in 12 hours) falling on large areas of recently drilled and bare ground (Boardman 1988): considerable damage to property occurred (Robinson and Blackman 1990, Table 3.2) and many fields suffered severe erosion.

Surveys of erosion in Britain show that it can occur on relatively low **slope gradients**. Evans (1980) for example, has shown that erosion can occur on slopes as low as 2–3°. The effect of **land management** on the occurrence of erosion events is, possibly the single most important factor in explaining why off-site impacts have apparently increased over the last decade or so (Figure 3.2). Land is more susceptible to erosion under some crops than others because of different planting dates in respect of rainfall, different row spacing, growth rates and therefore, effective ground protection against rainsplash. A figure of around

Table 3.5 Rainfall levels associated with selected cases of property damage due to runoff from agricultural fields.

Location	Date	Rainfall (mm)
Shepherds Mead, Worthing[1]	13.11.87	21.1
Pyecombe Golf Course[2]	26 & 27.10.90	27 & 38
Bishopsteignton, Devon[3]	20.12.89	50
Teffont, Wilts[4]	19 & 20.09.84	30
Steepdown, West Sussex[5,6]	26.10.90	22.6
Ashford, Derbyshire[7]	22.4.83	6

Sources: [1]Boardman, 1988a; [2]Boardman, 1990b; [3]Parkinson & Boardman, 1990; [4]Boardman, 1992; [5]Boardman and Evans, 1991; [6]Boardman and Evans, 1993; [7]Boardman and Spivey, 1987.

30% is frequently quoted as a cover value at which erosion significantly diminishes (e.g. Boardman, 1992a). Certain management practices associated with particular crops may encourage erosion; the frequency of vehicle traffic producing compaction is a problem on sugar beet and vegetable crops, as is the rolling of seedbeds on winter cereals and maize. Irrigation of salad and vegetable crops in Kent gives rise to high rates of erosion (Boardman and Hazelden, 1986; Evans, 1989). There are also frequently occurring associations of crops and soils, for example, sugar beet with sandy soils which influence erosion rates. Thus, it is difficult to separate crop characteristics, management practices and site properties in an assessment of erosion rates.

There is general agreement that the reported increase in significant lowland erosion events is due to the adoption of winter cereals and the consequent expansion of the area left bare in autumn and winter (Boardman and Robinson, 1985; Evans and Cook, 1986; Speirs and Frost, 1987). Indeed, the area sown to winter cereals has increased by over three times since 1969. Other factors that are believed to have contributed to the increase include:

- arable farming on steep slopes;
- the removal of field boundaries to create larger fields (Evans and Cook, 1986);
- inappropriate choice of crop on steep slopes, erodible soils etc.;
- working land up and down the line of maximum slope;
- presence of vehicle wheel tracks ('wheelings') which act as channels for runoff (Reed, 1986);

- rolling of seedbeds (Speirs and Frost, 1987).

The interaction between the various land management and physical factors results in **risk periods** associated with different crops defined by the bare ground associated with the growing of a particular crop and rainfall. Each of the six periods of erosion risk identified by Boardman (1991) is associated with a different land use or management practice:

(i) in late summer and early autumn in fields sown with oil seed rape or grass ley;

(ii) a short period in autumn when land has been cultivated before drilling winter cereals;

(iii) land drilled with winter cereals. The period of risk depends on rainfall distribution, drilling date and growth rate of the crop. In some years crop growth may be insufficient to inhibit erosion until April; in others the period of risk may extend only to December;

(iv) land ploughed and cultivated over winter and spring before sowing spring cereals;

(v) a short period of risk in spring on land drilling with spring cereals, although rapid crop growth can limit the risk period;

(vi) in May and June for land planted with late spring crops such as maize.

A similar situation can be identified in forestry practice, with the risk period corresponding to the

5–10 year interval between ploughing and the establishment of an effective tree cover or at the end of clearfelling and timber extraction.

The Identification of Vulnerable Areas

It is by no means straightforward to identify areas that may be at risk from soil erosion because of the strong influence of land use and management. In addition, there is a variety data sources that may provide an indication of vulnerable areas (Table 3.6). Possibly the most up-to-date map was produced by Evans (1990a) who classified the 296 soil associations of the National Soil Maps into five categories of erosion risk, including upland peat areas. Those categories are based on land use, landform and soil properties and take account of existing information on erosion in each association (Table 3.7). Evans (1990a) estimated that 6% of land in England and Wales is at high to very high risk of erosion.

Evans' classification together with information derived from the soil memoirs in Scotland, has been used to define erosion risk areas on the accompanying 1:625,000 scale thematic maps. However, it is important to note that 'at risk of erosion' does not mean that erosion occurs every year, only that it occurs regularly in some fields under current land use, management and climatic conditions. Indeed, a considerable area is potentially at risk; the fact that this potential is not fully realised is due to the fact that the present climate is not conducive to widespread erosion. However, land degradation is expected to increase as a result of global warming (Table 3.7).

Figure 2.1 provides an indication of the pattern of rainfall erosivity, highlighting the concentrations of higher values in parts of Upland Britain and throughout eastern and southern England. Erosivity is only a measure of the potential for erosion; actual erosion depends on the local conditions of soils, slopes and land management. Indeed, erosion is generally very localised in its occurrence, often being restricted to particular fields or even parts of fields. The distribution of erosion prone soils on the accompanying 1:250,000 scale maps tends, therefore, to overstate the real nature of erosion-related problems. However, in general terms, the most vulnerable areas include:

- the Lower Greensand soils of southern England and the Isle of Wight;

- sandy and loamy soils in Nottinghamshire, the West Midlands, Somerset, Dorset and parts of East Anglia;
- chalky soils of the South Downs, Cambridgeshire, the Yorkshire and Lincolnshire Wolds, Hampshire and Wiltshire;
- parts of eastern Scotland;
- upland peaty soils.

This list highlights those areas where, under particular circumstances, erosion could occur. In most cases this would only lead to hillslope or on-farm impacts. Significant off-site problems will only arise where:

- the sediment delivery from hillslopes to watercourses is high, leading to water pollution and sedimentation problems;

- development and infrastructure is placed at the foot of erosion prone slopes or in dry valleys where runoff may be concentrated (Figure 3.3).

It is important to bear in mind, however, that soil erosion frequently is associated with poor land management. Under appropriate management with a good plant cover, none of the soil associations identified as erosion prone would be prone to significant erosion.

The Effects of Development

Although there is often an increase in hillslope erosion during the construction period, built development generally results in a decrease in hillslope erosion risk because the ground is "protected" by hard surfaces. However, uncontrolled runoff from developments may lead to localised erosion problems downslope. This is not a frequently cited problem inland, but may contribute to cliff erosion on the coast (see Chapter 10). Possibly of greater concern is the way in which development has encroached into vulnerable settings such as dry valleys, transforming agricultural problems into significant hazards for some urban communities. Unfortunately potential problems frequently only become apparent after development has taken place and when there are changes in land management. It is not that problems are not foreseeable, rather that appropriate advice has not been sought or taken account of when planning new developments.

Table 3.6 Sources of information: Erosion prone soils.

i.	the 1:250,000 National Soil Maps of England and Wales and accompanying legend (Mackney et al., 1983) prepared by the Soil Survey of England and Wales (now the Soil Survey and Land Research Centre) allows soil associations at risk from erosion to be identifies. These maps have been summarised in various publications (Parry and others, 1991).
ii.	the 1:250,000 scale Soil Maps of Scotland and accompanying memoirs prepared by the Macaulay Land Use Research Institute. The maps and legend make no specific reference to erosion risk although information about each of the 558 map units can be derived from the memoir.
iii.	published data from the soil erosion monitoring scheme funded by the Ministry of Agriculture, Fisheries and Food and carried out by the then Soil Survey of England and Wales between 1982–86 covering seventeen sample locations (see Evans 1988, 1990a, b; Chambers et al., 1992).
iv.	the Soil Survey and Land Research Centre (formerly SSEW) have prepared 1:250,000 scale maps of erosion risk on **arable land**, published in 1994.

The Significance for Conservation

Areas of bare peat and gully systems on upland areas such as the southern Pennines have been frequently regarded as ecologically and aesthetically unacceptable. Erosion, in this context, is largely viewed as wholly negative and studies of the causes and treatment of upland erosion have been funded by many interested organisations, including the Peak Park Planning Board (Phillips et al, 1981).

The impact of soil erosion on fresh water streams and reservoirs had been examined earlier, although

Figure 3.3 Vulnerable settings for development in erosion risk areas.

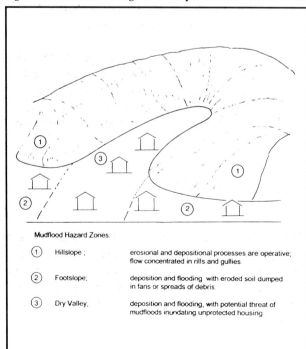

Mudflood Hazard Zones

① Hillslope ; erosional and depositional processes are operative; flow concentrated in rills and gullies.

② Footslope ; deposition and flooding with eroded soil dumped in fans or spreads of debris.

③ Dry Valley ; deposition and flooding, with potential threat of mudfloods inundating unprotected housing

the significance to conservation interests and natural habitats has received little attention in Britain. Evidence does suggest, however, that breeding fish in gravel bed rivers have been inhibited in recent years by sedimentation (see above for the possible impact of afforestation).

In general, hillslope erosion does not appear to be viewed as of major significance to conservation interests in Britain, with the exception of the degradation of upland peat landscapes. The volume of sediment moved on many hillslopes is generally limited; the supply to freshwater streams and rivers does not appear to generate major concerns over its impact on fluvial habitats. Other aspects of land management are probably more significant, most notably the supply of phosphates, nitrates and pesticides to streams from agricultural land.

Summary : The Significance to Planning and Development

Hillslope erosion and sediment–laden runoff is a frequent but, fortunately, relatively minor problem in many rural areas of Great Britain with erosion leading to land degradation or loss of soil productivity. However, in some rural areas the processes can result in notable local difficulties, or make a significant contribution to problems elsewhere in a catchment. Indeed, around 15% of soils in England and Wales are at moderate to very high risk of soil erosion (Table 3.7).

● erosion and runoff can create damage and inconvenience to communities in vulnerable areas, especially through the costs of clean–up operations. For example,

Table 3.7 Categories of erosion risk in England and Wales (after Evans, 1990a).

Category	Description	Actual Area %	Potential Area %*
Very small risk	Erosion occurs rarely or not at all.	38	28
Small risk	Eroding fields to cover less than 1% of the land each year.	38	26
Moderate risk	For arable land, eroding field cover between 1-5% of the land each year.	18	30
High risk	Erosion generally affects more than 5% of fields per year, and median and mean volumes eroded are likely to be greater than in the small risk categories.	4	13
Very high risk	In lowlands erosion rarely affects less than 5% of the fields each year. On average more than 10% of fields are affected and two years in five as much as 20-25% is affected. Volumes of soil eroded are greater than in other categories.	2	3

* Note: Estimated changes in erosion risk resulting from the effects of global warming.

the cost of the October 1987 erosion and mudflood events in the South Downs has been estimated to be in excess of £750,000 of which less than £15,000 was borne directly by the farmers (Robinson and Blackman, 1990);

- sediment carried into watercourses can lead to significant downstream problems for water supply companies (water quality and reservoir sedimentation), drainage authorities (increased flood risk), fisheries interests (water quality) and navigation and harbour authorities (decreased channel depth).

Changes to land use and management practices are considered to be a major factor in the apparent increase in significant events over the last 25 years. The increase in land sown for winter cereals or ploughed for commercial forestry are considered to be particularly important in increasing erosion risk and sediment yield from some hillslopes. It is anticipated that global warming may result in greater potential for erosion-related problems in many areas of the country (Parry and others, 1991; Table 3.7).

Although many aspects of land management are beyond planning control, local planning authorities can help ensure that the potential off-site problems

are minimised by ensuring that:

- development is not placed in vulnerable locations including dry valleys or potential mudflood tracks without proper precautions, such as the use of check-dams or runoff storage ponds;

- erosion prone soils are not exposed in construction sites at particular times of year, as this may lead to the generation of significant off-site problems;

- the potential effects of changes in land use on sediment yield could be an important planning consideration in determining planning applications or when reviewing environmental assessments for afforestation schemes.

Chapter 3 : References

Boardman J. 1983. Soil erosion at Albourne, West Sussex, England. Applied Geography 3, 317-329.
Boardman J. 1988a. Public policy and soil erosion in Britain. In J.M. Hooke (ed) Geomorphology in Environmental Planning, 33-50. John Wiley and Sons.

Boardman J. 1988b. Severe erosion on agricultural land in East Sussex, UK October 1987. Soil Technology 1, 33–348.

Boardman J. 1988c. Flooding and erosion at Shepherds Mead, Worthing. Report for Worthing District Council.

Boardman J. 1990a. Flooding at Pyecombe Golf Course, October 1990. Report for Pyecombe Golf Club.

Boardman J. 1990b. Soil Erosion in Britain: Costs, Attitudes and Policies, Social Audit Paper No. 1, Education Network for Environment and Development, University of Sussex.

Boardman J. 1990c. Soil erosion on the South Downs: a review. In, Boardman J., Foster I.D.L. and Dearing J.A. (eds), Soil Erosion on Agricultural Land, 87–105. John Wiley and Sons.

Boardman J, 1991. Land use, rainfall and erosion risk on the South Downs. Soil Use and Management 7, 34–38.

Boardman J. 1992a. Agriculture and erosion in Britain. Geography Review 6, 15–19.

Boardman J. 1992b. Effects of Agricultural Soil Erosion upon Watercourses: the River Allen Catchment, Report for Institute of Hydrology.

Boardman J. 1993. The sensitivity of Downland arable land to erosion by water. In, Thomas D.S.G. and Allison R.J. (eds) Landscape Sensitivity 211–228. John Wiley and Sons.

Boardman J. and Evans R. 1991. Flooding at Steepdown. Report for Adur District Council.

Boardman J. and Evans R 1993. Erosion and flooding risk, Steepdown 1992–93. Report for Adur District Council.

Boardman J. and Hazelden J. 1986. Examples of erosion on brickearth soils in east Kent. Soil Use and Management 2, 105–108.

Boardman J. and Robinson D.A. 1985. Soil erosion, climatic vagary and agricultural change on the Downs around Lewes and Brighton, Autumn 1982. Applied Geography, 5, 243–258.

Boardman J. and Spivey D.1987. Flooding and erosion in west Derbyshire, April 1983. East Midlands Geographer 10, 3–44.

Boardman J., Stammers R.L. and Chestney D. 1983. Flooding problems at Highdown, Lewes: technical report. Report to Lewes District Council.

Bower M.M. 1960a. Peat erosion in the Pennines. Advances in Science., 16, 323–331.

Bower M.M. 1960b. The erosion of blanket peat in the Southern Pennines. East Midlands Geographer. 13, 22–23.

Bower M.M. 1962. The cause of erosion in blanket peat bogs. Scottish Geographical Magazine 78 33–43.

Butcher D.P., Claydon J., Labadz J.C., Pattinson V.A., Potter A.W.R. and White P.

1992. Reservoir sedimentation and colour problems in southern Pennine reservoirs. Journal Institution of Water and Environmental Management 6, 418–431.

Chambers B.J., Davies D.B. and Holmes S. 1992. Monitoring of water erosion on arable farms in England and Wales, 1989–1990. Soil Use and Management 8, 163–170.

Cooke R U and Doornkamp J C 1990. Geomorphology in Environmental Management. Oxford University Press.

Crisp D.T. 1966. Input and output of minerals for an area of Pennine moorland: the importance of precipitation, drainage, peat erosion and animals. Journal Applied Ecology 3, 327–348.

Drakeford T. 1979. Report of survey of the afforested spawning grounds of the Fleet catchment. Forestry Commission.

Drakeford T. 1982. Management of upland streams. Inst. of Fisheries Management, 12th Annual Course, Durham.

Edwards A., Martin D. and Mitchell G. (eds) 1987. Colour in upland waters. Proc. workshop held at Yorkshire Water, Leeds.

Evans R. 1980. Characteristics of water–eroded fields in lowland England. In, De Boodt, M. and Gabriels D. (eds), Assessment of Erosion, 77–87. John Wiley and Sons,

Evans R. 1988. Water erosion in England and Wales 1982–1984. Report for Soil Survey and Land Research Centre, Silsoe.

Evans R. 1990. Soils at risk of accelerated erosion in England and Wales. Soil Use and Management 6, 125–131.

Evans R. 1990b. Water erosion in British farmers' fields – some causes, impacts and predictions. Progress in Physical Geography 14 199–219.

Evans R. and Cook S,. 1986. Soil erosion in Britain. SEESOIL 3, 28–58.

Ferguson R.I. and Stott T.A. 1987. Forestry effects on suspended sediment and bedload yields in the Balquidder catchments, Central Scotland. Trans. Royal Society Edinburgh, Earth Sciences 78, 379–384.

Forestry Commission 1988. Forests and water: Guidelines. Forestry Commission, Edinburgh.

Francis I.S. and Taylor J.A. 1989. The effect of forestry drainage operations on upland sediment yields: a study of two peat covered catchments. Earth Surface Processes and Landforms 14, 73–83.

Frost C.A. and Speirs R.B. 1984. Water erosion of soils in south–east Scotland – a case study. Research and Development in Agriculture 1, 145–152.

Labadz J.C., Butcher D.P. and Potter A.W.R. 1991. Moorland erosion in the southern Pennines Part One, Research Monograph No. 1, Department of Geographical and Environmental Sciences, The Polytechnic of Huddersfield.

Leeks G.J.L. and Roberts G. 1987. The effects of forestry on upland streams – with special reference to water quality and sediment transport. In J. Goode (ed) Environmental effects of plantation forestry in Wales. ITE Symposium 22 9–25.

Mackney D., Hodgson J.M., Hollis J.M. and Staines S.J. 1983. Legend for the 1:250,000 Soil Map of England and Wales. Soil Survey of England and Wales, Harpenden.

Moffat A.J. 1988. Forestry and soil erosion in Britain – a review. Soil Use and Management 4, 41–44.

Morgan R.P.C. 1980. Soil erosion and conservation in Britain. Progress in Physical Geography 4, 24–47.

Morgan R.P.C 1985. Assessment of soil erosion risk in England and Wales. Soil Use and Management 1, 127–131.

Morgan R.P.C. 1992. Soil conservation options for the UK. Soil Use and Management 8, 176–180.

Morgan R.P.C., Martin L. and Noble C. 1987. Soil erosion in the UK: a case study from mid–Bedfordshire. Silsoe College Occasional Paper 14.

Parkinson R.J. and Boardman J. 1990. Soil erosion and flooding at Teign View Road, Bishopsteignton, Devon 1989/90. Report for Messrs Jose, Kaye and Proctor.

Parry M.L. and others 1991. The Potential Effects of Climate Change in the United Kingdom, First Report, United Kingdom Climatic Change Impacts Review Group, HMSO.

Phillips J., Yalden D.W and Tallis J.H. 1981. Moorland erosion study. Phase I Report. Peak Park Joint Planning Board, Bakewell.

Quine T.A. and Walling D.E. 1991. Rates of soil erosion on arable fields in Britain: quantitative data from Caesium – 137 measurements. Soil Use and Management 7, 169–176.

Reed A.H. 1986. Soil loss from tractor wheelings. Soil and Water 14, 12–14.

Robinson D.A. and Blackman J.D. 1990. Some costs and consequences of soil erosion and flooding around Brighton and Hove, autumn 1987. In Boardman J., Foster I.D.K. and Dearing J.A. (eds), Soil Erosion on Agricultural Land, 369–382. John Wiley and Sons.

Robinson D.A., Williams R.B.G., Funnell D.C., Blackman J.D., Potts A.S., Browne T.J. and Boardman J. 1987. Flooding and soil erosion at Rottingdean, October 1987: final report. Report to Brighton Borough Council.

Soutar R.G. 1989. Afforestation and sediment yields in British fresh waters. Soil Use and Management 5, 82–86.

Speirs R.B. and Frost C.A. 1987. Soil water erosion on arable land in the United Kingdom. Research and Development in Agriculture 4, 1–11.

Stammers R.L. and Boardman J. 1984. Soil erosion and flooding on Downland areas. The Surveyor 164, 8–11.

Tallis J.H. 1964. Studies on Southern Pennine peats. 1–III. Journal of Ecology, 323–353.

Tallis J.H. 1986. Studies on southern Pennine peats. IV. Evidence of recent erosion. Journal of Ecology 53, 509–520.

Chapter 3: Suggested Reading

Boardman J., Foster I.D.L. and Dearing J.A. (eds) 1990. Soil erosion on agricultural land. Wiley.

Hodges R.D. and Arden–Clarke C. 1986. Soil erosion in Britain. The Soil Association.

Morgan R.P.C. 1980. Soil erosion. Longman.

4 Hillslopes: Soil blows

The Nature of the Problems

Strong winds blowing across exposed arable fields can erode soil particles and carry them long distances before depositing them in sheets of dust. The process depends on the availability of particles that can be picked up and wind speeds capable of moving them in suspension (high in the air), by surface creep (rolling on the ground) and saltation (bouncing). Little information is available on rates of wind erosion, although Wilson and Cooke (1980) indicate that between 2–4kg/m^2 may be lost each year in parts of the Vale of York.

Soil erosion by wind can damage crops and reduce productivity by removing seeds, exposing roots and blasting leaves, and by reducing soil quality (Cooke and Doornkamp, 1990). The dust storms can reduce visibility and block roads, ditches and fences. Although many farms in Britain may be affected by a degree of wind erosion with its cumulative effect on soil quality, notable off–farm problems tend to arise only in unusual circumstances. Documented examples of serious dust storms are rare, and the process rarely presents a significant constraint to land use and development. The following examples serve to illustrate how the process is essentially an agricultural issue.

Wind erosion is a regular phenomenon in east Shropshire, with a fairly severe blow every 3–4 years. Between 1967–1976, 1450 ha was affected (Reed, 1979) mainly in three parishes: Claverley, Rudge and Worfield. The most recent notable events took place in the spring of 1983, coinciding with a period of widespread wind erosion in the West Midlands (Fullen 1985). The events, between March and May, removed considerable quantities of topsoil; with around 6–10t/ha lost on an exposed, west facing field east of Bridgnorth. Stripped topsoil was deposited at the margins of

the eroded fields and on roads, forming small dune features 0.1–0.7m deep, comprising mainly fine sand. Very fine particles were transported beyond the small dunes and settled as a fine film of dust, covering exposed vegetation and road surfaces.

In March 1968 a series of soil blows across parts of Lincolnshire (and now Humberside; Figure 4.1) led to a range of local impacts (Robinson, 1969; Cooke and Doornkamp, 1990). Many roads were partially blocked by wind–blown material and traffic was disrupted. Clearing operations in Lindsey alone cost £4000. Ditches and drains were filled with sediment: it cost one drainage board £5000 to clear ten drains. Many farmers had to clear ditches on their own land, at their own expense: the average cost was estimated to be approximately £5 per 20 metres, and in the Isle of Axholme alone the cost may have been £17,500. Also on farmland, productivity was reduced in places by uncovering or removal of seeds (e.g. barley, peas, and beet), and by the 'scorching' of leaves and root–exposure of winter wheat plants.

Severe wind erosion can also be a problem in the Fens, especially in areas of reclaimed fen peat and sandy soils. Most years in spring, the fine tilth of the peaty soils is susceptible to wind erosion; seed, fertiliser and up to 2cm of topsoil can be blown away and dykes infilled. Where damage is severe, crops have to be resown (Hodge et al, 1984). Unless the dykes are cleared out, flood problems can occur when the events are followed by heavy rain (Pollar and Millar, 1968; Perry, 1981).

East Yorkshire has also had a long history of localised wind erosion, leading to temporary sheets of sand on roads and dust in houses. The largest recent events occurred during February and March 1967 within the "Great Sand Field" around Market Weighton (Radley and Simms, 1967). Gale force winds picked up the fine sandy soils and deposited them in sand dunes and sand sheets. At Moxby,

Figure 4.1 Wind erosion in Lincolnshire, 1968 (after Robinson, 1969).

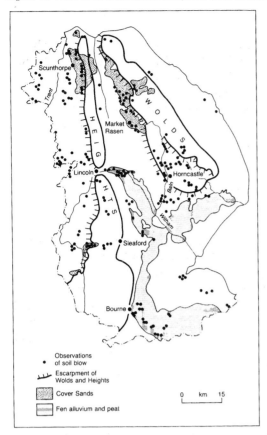

dust storms restricted visibility for more than a mile. The York–Helmsley road was blocked, and the road was only kept open by a road sweeping vehicle working continuously for a week. Dunes over 1m high were visible for several weeks after the blows.

Perhaps the most remarkable inland soil blows in Britain occurred during exceptionally severe storms between 1570 and around 1668, in the Breckland of East Anglia. The town of Santon Downham was gradually engulfed by moving sands by around 1630; farmhouses were buried and later exposed as the sands moved on. Attempts to slow down the sand movement produced "sandbanks near 20 yards high" (Wright, 1668). The river at Santon was nearly blocked for 5km. Lamb (1991) suggests that between 50–100Mm^3 of sand was involved in the "sand floud" or "wandering sands", possibly derived from exposed sandy soils around Lakenheath. The famous diarist John Evelyn described the scene in 1677:

"The Travelling Sands .... that have so damaged the country, rouling from place to place ... like the sands in the Deserts of Libya, quite overwhelmed some gentlemen's whole estates" (John Evelyn, 1677).

The area has since been stabilised by afforestation, but the events highlight the scale of destruction that can occur in the severest of storms, if sandy soils are left exposed.

Causes of Wind Erosion

Occurrence of significant wind erosion events generally corresponds with the combination of:

- **erodible** soils;
- strong, **erosive** winds;
- the **land management** practices.

The most **erodible** materials tend to be fine sandy soils and lowland peats. Morgan (1985) suggests that the whole of Britain experiences wind speeds greater than a threshold erosivity level at least once a year (this level is taken as 34km/hr or around 17 knots and wind force 5; see Figure 4.2). However, **land management** practice is the dominant control in determining the occurrences and timing of erosion events. Soils tend to be at greatest risk when they are left bare with a fine, smooth surface in spring and early summer (Armbrush et al. 1964); removal of hedgerows increases the erosivity of the wind by reducing the number of wind breaks.

Indeed, changes in land management since World War II have tended to increase the hazard (Table 4.1; Wilson and Cooke, 1980). It seems probable that these changes led to an increased susceptibility to wind erosion, and when climatic conditions are appropriate, soil blowing may be serious. This was the case in Lincolnshire during March 1968 when three climatic factors combined to produce favourable conditions for severe erosion:

(i)   precipitation, in the first three months of the year, was only 54% of the average: the soils were abnormally dry;

(ii)   the number of frosts in February was significantly higher than average in some areas: frost action may have helped to produce a finer soil tilth than usual;

(iii) there was a period in mid–March of very strong westerly winds associated with a series of vigorous low–pressure troughs which followed the dry, frosty winter and preceded the growth of crops.

It should be recognised, however, that trends towards increasing susceptibility to wind erosion

Figure 4.2 Average wind directions and frequencies around Great Britain (after Chandler and Gregory, 1976).

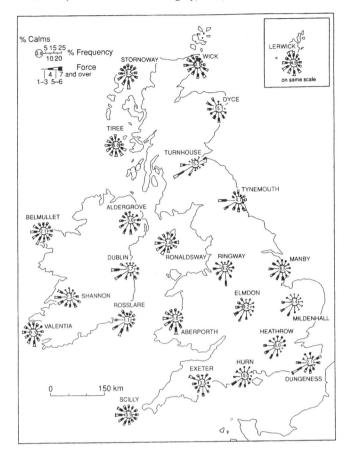

Figure 4.2 Average wind directions and frequencies around Great Britain (after Chandler and Gregory, 1976).

The Effects of Development

Severe soil blows and dust storms are essentially an agricultural problem. The encroachment of built development into areas vulnerable to wind erosion will result in a dramatic reduction in the supply of material, although the new buildings and roads themselves may be affected by the transport of eroded material from elsewhere.

The Significance for Conservation

There is no evidence to suggest that wind blown dust from agricultural land can be considered as anything more than a temporary nuisance to conservation interests. No habitats or landforms are dependent upon the supply of wind borne sediment, none suffer lasting damage. This, of course, is in direct contrast to wind erosion in sand dunes which is essential for maintaining the dynamic nature of the landforms and the habitats which they support (see Chapter 11).

Summary: The Significance to Planning and Development

Soil blows and dust storms can be occasional minor problems in some rural areas, especially in parts of lowland England and Wales. Although over 6000km² of sandy and silty soils and lowland peats are considered to be prone to wind erosion and the whole country experiences winds above a critical erosive threshold (force 5) at least once a year, the realisation of this potential is usually limited to arable farms when they are left bare in the spring and early summer.

In general, the most frequent problems are related to land degradation or loss of productivity in affected fields. However, locally significant off-site problems may occur in the more severe events, including:

- **traffic disruption** through blocked roads and reduced visibility;

- **increased flood risk** where watercourses or drainage channels have been infilled by wind-blown sediment. Flood problems can occur when soil blows are followed by heavy rains.

have begun to be reversed as affected farmers have adopted new techniques to limit damage.

The Identification of Vulnerable Areas

The most up-to-date map of areas prone to wind erosion was prepared by Evans (1990) which, together with information derived from soil memoirs in Scotland, forms the basis for identification of erosion risk areas on the accompanying 1:625,000 scale thematic maps. Over 6000km² in England and Wales is reported to be prone to wind erosion (Morgan 1985). The most severe wind erosion problems are associated with:

- peaty soils, such as in the Fens, western Lancashire and the Somerset Levels;

- sandy soils, frequently corresponding with fluvio-glacial outwash deposits, as in the Vale of York, Lincolnshire, the Brecklands of East Anglia, the East Midlands and East Shropshire.

Table 4.1 Increased wind erosion in Britain since World War II: Some suggested causes and their consequences (after Wilson and Cooke, 1980).

1.	An increase in arable land and resulting reduction in permanent pasture and the use and length of grass leys, which has reduced the protective effect of vegetation, and extended both the areas susceptible to wind erosion and the periods of susceptible to wind erosion and the periods of susceptibility (arable land has recently decreased due to set-aside policies).
2.	A tendency towards monoculture, and a consequent reduction of the stability provided by crop rotations.
3.	The practice of stubble and straw burning (now banned) which reduces the protective effect of such material and reduces the provision of organic matter to the soil of value in maintaining soil structure.
4.	An increase in the use of artificial fertilizers, some of which tend to disaggregate soil clods.
5.	Improved weed control with herbicides, which reduces the protective effect of weeds.
6.	The introduction and rapid extension of sugar beet (now declining), a crop requiring a loose, vulnerable seed bed and providing only limited surface protection early in its growth season.
7.	An increase in the use of wide-spaced 'drill-to-stand' techniques which increase the area of ground vulnerable to wind erosion in a field.
8.	Removal of hedges and increase in field size, thus reducing the protective effect of field boundaries and increasing wind 'fetch'.

Note: Recent changes in agriculture policy and land management practice have begun to reverse the trend towards increased susceptibility to wind erosion.

Soil blows are principally a land management problem rather than planning and development issues. However, in erosion prone areas it may be prudent to minimise the removal of wind break features such as hedgerows or lines of trees during development. In such areas the potential for partial infilling of watercourses may need to be taken into account in the design of site drainage measures and in the imposition of planning conditions to ensure access for removing sediment after an event.

Chapter 4 : References

Armbrush D.V, Chepil W.S. and Siddoway F.H. 1964. Effects of ridges on erosion of soil by wind. Soil Science Society America Proc. 28, 557–560.

Chandler T J and Gregory S 1976. The climate of the British Isles. Longman.

Cooke R.U. and Doornkamp J.C. 1990. Geomorphology in environmental management. Oxford University Press.

Evans R. 1990. Soils at risk of accelerated erosion in England and Wales. Soil Use and Management 6, 125–131.

Evelyn J. 1668. Diary of John Evelyn.

Fullen M.A. 1985. Wind erosion of arable soils in East Shropshire (England) during spring 1983. Catena 12, 111–120

Hodge C.O., Burton R.G.O., Corbett W.M., Evans R. and Seale R.S., 1984. Soils and their use in Eastern England. Soil Survey of England and Wales Bulletin No. 13.

Lamb H.H. 1991. Historic storms of the North Sea, British Isles and Northwest Europe. Cambridge University Press.

Morgan R.P.C. 1985. Assessment of soil erosion in England and Wales. Soil Use and Management 1, 127–131.

Perry A.H. 1981. Environmental hazards in the British Isles. George Allen and Unwin.

Pollard E. and Millar A. 1986. Wind erosion in the East Anglian Fens. Weather 23, 415–417.

Radley J. and Simms C. 1967. Wind erosion in East Yorkshire. Nature 216, 20–22.

Reed A.H. 1979. Accelerated erosion of arable soils in the United Kingdom by rainfall and runoff. Outlook on Agriculture 10, 41–48.

Robinson D.N. 1969. Soil erosion by wind in Lincolnshire, March 1968. East Midlands Geographer 4, 351–262.

Wilson S.J. and Cooke R.U. 1980. Wind erosion. In M.J. Kirkby and R.P.C. Morgan (eds) Soil erosion, 217–251. Wiley.

Wright T. 1668. Philosophical Transactions of the Royal Society. Vol III No. 37, 722–725.

Chapter 4 : Suggested Reading

Morgan R.P.C. 1980. Soil erosion. Longman.

5 Rivers: Flash floods

The Nature of the Problems

Flash floods are characterised by the rapid rise and fall of the floodwaters, with peak flows often occurring within hours of the onset of heavy rain. Although they are frequently associated with upland and mountainous areas, their occurrence reflects the regime characteristics of a particular river or stream (Chapter 2). Thus, flash floods can occur in many catchments throughout Britain from mountain rivers in the Scottish Highlands to winterbourne, seasonal streams in chalk areas (e.g. the Louth floods of May 1920 and Chichester floods of 1994).

It is, however, difficult to define what constitutes a flash flood as, in reality, there is a gradual progression from these events to seasonal flooding on major rivers. Perhaps the most important factors are the unexpected nature of the flooding, the unusually large damages that can be associated with the high discharges and velocity of flow, and the widespread erosion and deposition that frequently accompany such events.

These floods are often the result of extreme rainfall events; discharges may be exceptionally high and difficult to predict. Extreme rainfall events can occur during major thunderstorms or when fronts are stationary. Occasionally the two processes can combine and form an exceedingly rare occurrence such as the Lynmouth floods of August 1952, which produced a discharge of 511 m³/sec that had only been exceeded twice on the Thames (draining an area 100 times larger) since 1883. An intense summer storm of around 225mm fell over Exmoor and the steep Lyn catchment, one of the heaviest 24-hour totals ever recorded. The storm was related to a slow moving occluded front which developed into a thunderstorm as it intensified over Exmoor (Bleasdale and Douglas, 1952). The flood response was very rapid (Figure 5.1), reflecting, in

part, the preceding wet conditions in early August. A "wall of water" or "tidal wave" swept down the steep channel, as temporary dams formed by trees and boulders were breached (Marshall, 1952; Kidson, 1953). In the town of Lynmouth, a riverside holiday resort on the north Devon coast, there was widespread devastation; 25 adults and nine children were killed, 90 houses destroyed and 130 cars swept away. Local residents and holiday-makers were evacuated on the orders of the health authority, fearful of the spread of disease. The town was littered with debris left by the floodwaters, including boulders up to 7.5 tonnes found in the basement of the Lyn Valley Hotel. The total damage caused by the disaster has been estimated at £9M (Newson, 1975).

Similar, albeit less damaging, events have been recorded elsewhere in south west England (Table 5.1), often initiated by extremely intense rainfall. Indeed, it appears that the south west is particularly susceptible to heavy summer thunderstorms. The falls at Martinstown, Bruton and Cannington are the three largest daily rainfall events recorded in Great Britain; it can be no surprise that they caused dramatic flood events.

The Highlands of Scotland are also noted for frequent flash flooding (Table 5.2); in one event, in 1829, the River Findhorn rose 15m above its normal level during the course of a storm (Nairne, 1895). Such floods are frequently accompanied by widespread erosion and landsliding on the hillsides, as in the 1956 Cairngorm floods, and deposition of spreads of gravel and cobbles across the river floodplains. Although the Highland floods are probably amongst the largest in Britain, the predominantly rural nature of the area has fortunately prevented enormous damages. They do demonstrate, however, that flash floods can carry large quantities of debris, especially sand, gravel and boulders which can be moved downstream and left spread across the flooded area. For example,

Figure 5.1 The rainfall patterns associated with the 1952 Lynmouth flood (after Ward, 1978).

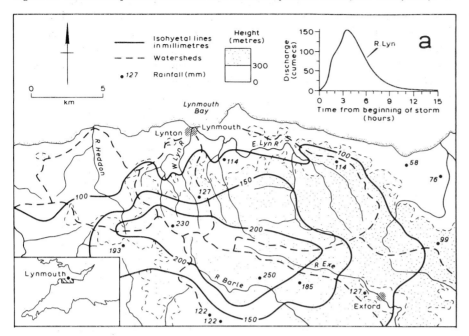

during the 1829 floods in the Moray and Inverness area, up to 2m of sand was deposited in valley floors, severely damaging agricultural land.

Flash floods have frequently occurred in the Pennine valleys. For example, on 8 August 1967 three houses were destroyed and ten more badly damaged in the village of Wray on the northern side of the Forest of Bowland, Lancashire. The Dunsop stream rose nearly 6m in 45 minutes during a thunderstorm with rainfall intensities of

around 117mm in 90 minutes. In Calderdale, West Yorkshire, flash floods on 19 May 1989 followed an extremely intense storm, believed to be over 193mm in two hours. Floodwaters destroyed a culvert adjacent to a row of terrace houses in Luddenden, leaving them on the edge of a 20m wide, 5m deep hole. In the centre of the village houses were flooded to a depth of more than a metre and several cars and a caravan were swept away as the river rose 3.5m in only 20 minutes (Acreman, 1989).

Table 5.1 Examples of flash flood events south-west England.

Date	Location	Comment
28–29 June 1917	Bruton, Somerset	242.8mm of rain fell on Bruton; town flooded, although not disastrously.
18 August 1924	Cannington, Somerset	238.8mm of rain, including hailstones up to 150mm diameter; locally severe flooding due to breaching of dam-blocks across the torrents.
15 August 1952	Lynmouth, Devon	Around 225mm of rain; severe flooding aided by "physiographic aggravation" due to extremely steep narrow catchment and valley of R. Lyn. Loss of life; 25 adults and 9 children. Disaster cost £9M.
18–19 July 1955	Martinstown, Dorset	297.4mm of rain, accepted as the largest rainfall day total in Great Britain. Bridges and gardens washed away at Osmington; flooding at Coryates, Upwey, Winfrith, Weymouth; up to 1.5m of water.
10 July 1968	Cheddar Gorge, Somerset	Around 170mm of rain; 6 dead. Houses flooded, extensive erosion.
27 January 1988	Truro, Cornwall	Around 90mm of rain; residential and city centre commercial properties flooded, traffic disrupted.
11 October 1988	Truro, Cornwall	Around 60mm of rain; extensive city centre flooding.

One of the most disastrous flash flood this century was on the margins of the Lincolnshire Wolds, at Louth on 29 May 1920. Up to 153mm of rain fell in the small chalk catchment, eroding fields and creating 18–30m wide torrents in normally dry valleys, probably similar to the mudfloods described in Chapter 3. The River Ludd rose 5m in 15 minutes and a 60m wide flood wave carrying about 140m³/sec swept through the small town of Louth. Twenty–two people were drowned, buildings demolished, 1,250 made homeless and over £100,000 of damage was caused. The Ludd was not known to have overflowed before.

The Great Till flood of 16 January 1841 also occurred in a chalk–dominated catchment, when a combination of melting snow and frozen ground resulted in a flash flood from the downlands of Salisbury Plain. The River Till burst its banks at Shrewton and rose to 2.5m above normal level. Three people drowned and 200 were left homeless,

many having had narrow escapes; one man even took his pigs upstairs with him for safety. The flood destroyed 72 houses and caused an estimated £10,000 worth of damage in the villages of Tilshead, Orcheston, Shrewton, Maddington and Winterbourne Stoke (Cross, 1967).

In January 1994, parts of Chichester and the surrounding countryside were flooded when the River Lavant, a minor seasonal chalk stream, rose dramatically after a prolonged wet period and a very intense rainstorm. Whereas in preceding years the river had hardly flowed at all, in early 1994 it carried 8.3m³/sec, nearly 20 times its normal flow for January (Taylor, 1994). The Hornet district of the city was inundated several times when a culverted section could not contain the high flows; sixteen "green goddess" reserve fire engines were called in to pump water to the harbour. Because of the flooding of the surrounding area, the army was needed to build two bailey bridges across the

Table 5.2 Examples of flash flood events in the Scottish Highlands (from Nairne, 1895 and later sources).

Date	Location	Comment
3–4 August 1829	Moray district	Around 90mm of rain; the R. Findhorn rose 15m above its normal level causing immense damage; bridges swept away, crops and farms destroyed or ruined by deposited gravels. Numerous families left destitute and damage estimated at £20,000 (1829 prices). Severe floods on the Nairn and Spey; "great landslips" occurred, farms swept away. In Spey valley damages estimated at over £37,000, plus countless livestock and several lives lost.
27 August 1829	Inverness district	Considerable flood damage in Inverness; crops flattened, numerous bridges lost, mills and homes damaged. Estimated as several thousand pounds damage.
24–26 January 1849	Inverness district	The "Inverness Flood"; most disastrous flood in the NW Highlands. Bridges lost at Aberchalder and Forst Augustus, Caledonian Canal breached. In Inverness the stone bridge was lost and a third of the town flooded by the combined waters of the Ness and the canal. Immense damage but no loss of life.
30 January – 1 February 1868	Inverness district	Farms damaged, crops lost, bridges swept away throughout district. Inverness flooded with extensive damage to property.
29 January 1892	Strathglass, Strathspey	Great, extensive flood following unpredicted snow falls for ten days. Damage extensive especially in Strathglass, Bonar–Bridge and Strathspey, but no fatalities. Railways washed away on Skye.
25 May 1953	Lochaber, Appin and Benderloch	Road bridges destroyed, disruption to road and rail traffic; extensive damage to forestry property through floods and landslips. In Argyllshire damage to roads estimated at £130,000.
30 July 1956	Cairngorm and Moray	Flooded houses and bridges damaged throughout region, especially around Forres. Main railway line from Inverness washed away. Livestock swept away. Extensive erosion and deposition of gravels on agricultural land. 72hr maximum rainfall of 250mm.
17–18 August 1970	Moray	72hr maximum rainfall of 150mm. Extensive damage to roads and bridges, agricultural land flooded and covered by gravels.

floodwaters to ease the traffic disruption on the main routes into the city.

In many cases the impact of flash floods may extend beyond the inundation of communities adjacent to the affected watercourses. Such events can also be very effective in mobilising sediment in upland areas and initiating both hillslope and channel erosion (Newson,1989). For example, in August 1986 Hurricane Charley caused widespread flooding in the Northern Pennines and was accompanied by channel incision to bedrock in many rivers, in-channel deposition of cobble and boulder fans and deposition of extensive sheets of fine grained sediments across floodplains for up to 75km downstream, as on the Lower Tyne (Newson and Macklin, 1990). Indeed, the extensive sediment transport associated with flash flooding can contribute to a number of significant downstream problems most notably:

- sedimentation of reservoirs and decline of water quality (see Chapter 3);

- river channel changes (see Chapter 6);

- increase in flood risk due to deposition within channels (see Chapter 7).

The Causes of Flash Flooding

The main regime characteristics controlling the occurrence of floods have been outlined in Chapter 2. However, amongst the most important factors influencing flash floods are:

- **catchment characteristics**
- **rainfall patterns**
- **land use and management**

Flash floods tend to be associated with short, relatively steep rivers and streams, frequently with average channel gradients exceeding 1:500 (2m/km; Figure 5.2). This includes steep **mountain streams** which have steep side-slopes with no floodplain separating the slopes from the channel and, hence, no significant storage capacity for floodwater or sediment, and relatively steep **hillside streams** with narrow floodplains adjacent to the active channels.

The catchments of many upland areas are generally small with steep valley-side slopes and are frequently underlain by impermeable Precambrian, Lower Palaeozoic and Carboniferous rocks (Figure 5.3). Only in areas of limestones is there significant

recharge of groundwater through deep percolation. The dominant catchment characteristics of steep slopes and impermeable subsoils leads to very high runoff, which may be up to 90% of rainfall.

The uplands are, of course, the wetter areas of Great Britain with the greater relief causing an intensification of rainfall, particularly in advance of cold fronts (Figure 2.13). As outlined in Chapter 2, the annual rainfall totals can be very high; 6530mm was recorded at Sprinkling Tarn in the Lake District in 1954. This rainfall is generally dominated by frequent but relatively low intensity storms associated with winter depressions, with upland areas having significantly more rain days than the lowlands (Figure 5.4). However, as the preceding discussion has highlighted, very intense storms do occur, albeit infrequently.

The combination of steep terrain, very high runoff, small catchments and episodic heavy rainfall creates the unique flash flood problems of upland areas. Rainfall intensities of around 6mm per hour are capable of creating spates (bankful conditions) in many upland streams, although much greater intensities are needed to generate memorable flash flood events. The rapid rise of floodwaters is a reflection of the speed at which water is delivered to the stream channel because of the very high runoff rates and the fact that water arrives at the channel at nearly the same time throughout the small catchment (i.e. a short **time of concentration**). The absence of significant floodplain areas adjacent to many upland streams tends to intensify the flood conditions as floodwaters are not "stored" in relatively slow-flowing backwaters.

Land use and management is believed to have an important role in determine the flood character of upland areas (see Chapter 3). The nature of the vegetation cover and land use is an important factor; peat areas tend to absorb rainfall and dampen the flood response; typical "moorland" vegetation of short grass, bracken or heather tends to have low evapotranspiration rates and, hence, the majority of rainfall is shed as runoff; upland afforestation tends to increase these evapotranspiration losses through the interception of up to 20% of the received rainfall (Calder & Newson, 1979), although this is unlikely to be significant in reducing the impact of extreme rainfall events.

There are also concerns that forestry operations may lead to increased flood problems, as was suggested by Howe et al., (1967) for the Severn

Figure 5.2 Average channel gradients in Great Britain (after Newson, 1978).

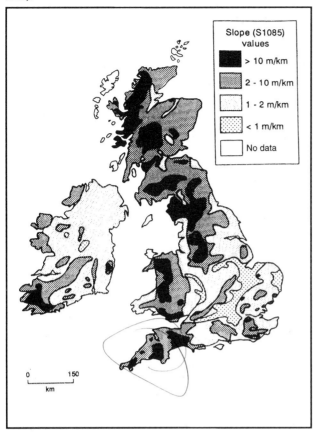

Slope (S1085) values

- ■ > 10 m/km
- ▨ 2 - 10 m/km
- ░ 1 - 2 m/km
- ▦ < 1 m/km
- □ No data

0   150
km

Figure 5.3 The ancient rocks of upland Britain.

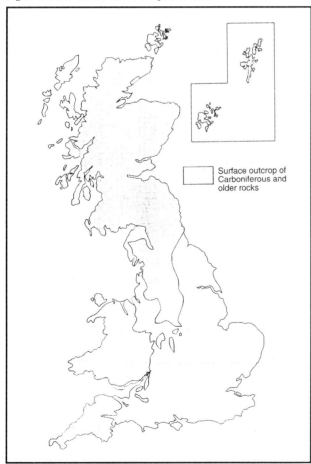

Surface outcrop of Carboniferous and older rocks

and Wye catchments. The extensive drainage operations necessary for planting in peaty areas can have immediate effects on slope hydrology by increasing runoff and flow in the upland streams, and reducing the time to reach peak flow. This effect is enhanced by the practice of ploughing downslope, creating numerous artificial channels. Robinson (1981) demonstrated that in the first seven years after ploughing a north Pennine catchment for afforestation, there were higher flood flows and the time to peak flow was halved. Whether these changes are temporary, becoming less significant as the forests mature and intercept more rainfall, is difficult to define. For example, Robinson and Newson (1986) noted that faster flood responses still occurred from a mature forested catchment at Plynliman than from open moorland. However, it should be noted that in many new forests, drainage ditches cease to run with water once the canopy is fully formed (Timber Growers Association, 1986). Acreman (1985) noted that the effects of afforestation on the flood hydrograph were dependent on the position within the catchment. Whereas afforestation downstream can lead to reduced peak flows, upstream afforestation can result in higher peak discharges, of up to 37%, and a decrease in the time of concentration.

Perhaps the greatest transformation of upland streams has been the construction of dams and reservoirs to provide water for use in the urban centres of lowland areas. These reservoirs can have a significant role in regulating the flood hazard by storing floodwaters and reducing peak flows. However, the structures themselves represent a potential flood hazard. Indeed, dam failures, although very rare, have resulted in a number of disastrous flash flood events; the most recent with loss of life was in 1925 (Table 5.3). More recent dam failures in this country, although causing major floods, have fortunately not caused loss of life.

The most disastrous failure in terms of damage and loss of life occurred in Yorkshire during March 1864 when the Dale Dyke Dam collapsed. The dam had been completed in 1863 and was almost full by March the following year. During the afternoon of March 11, a crack was observed along the downstream side. At 11.30 p.m. the same evening, the dam suddenly collapsed with no further warning. Almost 700 million gallons of water was released into the Loxley valley, a tributary of the River Don. Nearly 250 people were swept away as the flood destroyed the villages of Low Bradfield, Damflask and Little Matlock.

Further downstream, parts of Sheffield were completely demolished, with many buildings simply washed away with their occupants still inside. An inquest into the disaster held later that month suggested that the failure was due to seepage and erosion. The impact on the community was enormous. Over and above the tragic deaths, there were 7,300 compensation claims lodged by flood victims. The total cost to the water company in terms of compensation, repairs and fees, was more than £420,000 (1864 prices). This sum was raised by a 25% water levy over 25 years, and led to public outrage that the victims of the flood were left to foot the bill for the disaster.

In lowland catchments flash floods are frequently the product of a combination of high groundwater levels or frozen ground (which limit infiltration) and very intense storms. The Chichester floods of January 1994, for example, were associated with abnormally high groundwater levels in the Chalk downs, which had risen 27m in 41 days during the wet autumn and early winter of 1993 (Taylor, 1994). An intense storm on 30 December 1993 of around 48mm in four hours, could not be absorbed by the saturated ground leading to very high discharges in the River Lavant and flooding in Chichester. The river remained high over the following weeks because of the continuing supply of groundwater from springs around the chalk outcrop.

Flood problems can also result from the back-up of water behind blocked or under-sized culverts. The Truro floods of January and October 1988 (Table 5.1) were caused, in part, by floodwaters exceeding the capacity of city centre culverts. The two floods generated flows of 22.5m³/s and 31m³/s, respectively, considerably exceeding the culvert capacity of 15–18m³/s. Blocked culverts and temporary dams created by debris lodged behind bridge spans can also be an important factor in intensifying flood conditions. During the extensive flooding in South Wales on 27 December 1979, there were many examples of floods caused in this way (Vaughan, 1980), including:

- Nant Rhyd–Y–Car, near Merthyr; a culvert under a disused railway embankment was blocked and water ponded behind. The embankment failed, releasing the floodwaters which were trapped again behind a footbridge blocked with debris, flooding a row of houses.

- Blaen–Y–Cwm, at the head of the Rhondda Fawr; widespread erosion of

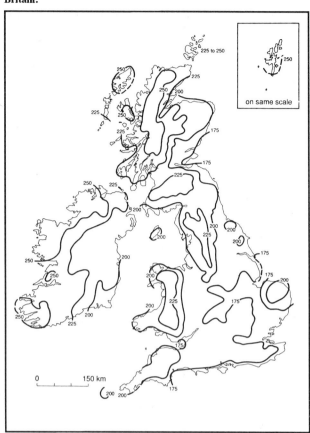

Figure 5.4 The average number of raindays per year in Great Britain.

stream banks led to the blocking of a culvert under the main road with 1500 tonnes of debris. Flooding occurred, badly damaging the road, cutting off the village and requiring 150 people to seek emergency accommodation.

- Ystradgynlais, in the Tawe valley; tree debris carried by the river formed a dam behind a bridge, causing floodwaters to build up and enter riverside houses.

The Identification of Vulnerable Areas

Flash floods are a characteristic of small steep catchments affected by episodic heavy rainfall. Although, the flood behaviour of individual catchments is extremely complex a **general indication** of flash flood potential can be derived from the mean channel slope within a catchment. The 21 catchments listed in Table 5.4 are the steepest in Britain and together with those areas with channel gradients in excess of 1:500 (Figure 5.2) must be considered prone to flash floods. However, many of the major events described in this Chapter lie outside these areas (especially the

Table 5.3 Fatal dam failures in Great Britain (after Charles, 1992).

Date	Site	Comment
1799	Tunnel End, Marsden	Overtopped during floods; one dead.
1810	Swellands, Huddersfield	Failure probably due to under seepage; Colne Valley flooded with five dead.
1835	Whinhill	Overtopped during floods; 31 dead.
1841	Welsh Harp	Two dead; overtopped during floods.
1842	Glanderston	Overtopped during floods; 8 dead.
1848	Darwen, Blackburn	Dam failure during heavy flood; 12–13 lives lost.
1852	Bilberry, Holmfirth	Settlement caused by internal erosion led to overtopping and collapse; 81 dead.
1863	Dale Dyke, Sheffield	Dam breached during first filling of reservoir; 244 lives lost and extensive property damage in Sheffield.
1870	Rishton	Three dead; unknown cause.
1875	Carne, Wales	Twelve dead; failure due to internal erosion.
1875	Castle Malgwyn	Two dead; overtopped during floods.
1910	Clydach Vale	Overtopped during floods; five dead.
1925	Skelmorlie, Largs, Scotland	Overtopped and breached during flood caused by release of water from flooded quarry; 6 dead.
1925	Coedty, Dolgarrog, Wales	Overtopped and breached during flood caused by collapse of concrete Eiglau Dam; 16 dead.

4 Flood survey

disastrous Louth floods), indicating that slope gradient is not necessarily the dominant factor in controlling their occurrence.

The area at risk within a vulnerable catchment is likely to be a narrow strip alongside stream channels or within dry valleys (Figure 5.5). Such areas can be readily defined from morphological evidence. It is also important to recognise the potential hazards within valleys below dams and reservoirs (Bossman–Aggrey et al, 1987).

Perhaps the greatest problem associated with flash floods is their unpredictability. The Truro floods of 1988 (Table 5.1) also highlight the difficulty in attempting to quantify the likelihood of rare flood events. Based on the historical record, the floods of 27 January were identified as having a return period of 350 years; thus most residents considered it would be unlikely that a similar event would occur in their lifetime. The 11 October floods were, however, larger with an estimated return period in excess of 400 years. The return periods were reassessed in the light of the most recent events to be 50 and 100 years, respectively (Acreman, 1988,

1990). Even so, the chance of getting consecutive 50 year and 100 years floods in the same year is 1 in 2500.

The 1988 Truro floods highlight the problems of assessing risk on the basis of return period statistics. Indeed, as stressed in Chapter 2 return periods may have been derived from very short periods of rainfall or flow records and will invariably be not only "best estimates" but subject to significant changes whenever a large flood occurs.

The Effects of Development

Many upland catchments remain largely undeveloped and, hence, the additional runoff generated from impervious surfaces may not significantly increase the of flash floods. However, development of an area can dramatically increase the runoff generated from rainfall events. This surface water is channelled through storm drainage and sewers into a nearby stream or river. However,

Table 5.4 Steep upland catchments and hydrometric areas in Great Britain (see Figure A.1 for location of areas).

Catchment		Channel Slope m/km	Area km²
Garry	(91)	117.7	1327
Fyne	(87)	94.6	720
North Gwynedd	(65)	51.5	1317
Aire	(90)	30.6	1177
South Devon	(46)	29.9	1512
Tees	(25)	26.7	2238
North Mersey	(69)	22.8	2687
Tay	(15)	19.5	5080
North Devon	(51)	17.8	528
Wear	(24)	15.8	1197
Wye	(55)	15.7	4184
Dumbartonshire	(85)	15.5	814
Taw and Torridge	(50)	14.9	2146
Exe	(45)	14.5	2253
Ribble	(71)	14.4	1488
Clwyd	(66)	14.2	1503
West Cumbria	(74)	13.8	914
Mid Glamorgan	(58)	12.6	1028
South Cornwall	(48)	12.2	1559
South Lakeland	(73)	12.2	1202
Eden	(76)	12.2	2397

Note: See Figure A.1 for location of catchments

in heavy storms the amount of runoff can exceed the capacity of the drainage system, leading to flooding. For example, an intense hail and rain thunderstorm in the Oxford area on the evening of 13 July 1967 led to extension flood problems. The heaviest rain fell in Cowley, flooding houses, roads and a number of factories including the British Motor Corporation. A new housing estate at Blackbird Leys suffered severe flooding with houses under 1m of water. It is estimated that over 70mm of rain fell in just over an hour (McFarlane and Smith, 1968). Flood problems were accentuated by heavy hail blocking the drainage system and some blocking of street drains and open culverts by garden rubbish.

Urbanisation can lead to an intensification of summer convection storms, making some areas more prone to flash flooding. Hampstead, for example, has been affected by seventeen flash flood events this century, including the dramatic floods of 14 August 1975 when around 160mm of rain fell in 2.5 hours. Streams were unable to cope with the runoff and were soon overtopped, sending large waves over garden walls and flooding over a hundred houses. One man drowned in his basement flat.

Most urban storm-drains are designed to cope with rainfall intensities of between 1–3 inches an hour, depending on the size of the area. However, rainfall events that could lead to such urban flooding are relatively common in Great Britain;

Figure 5.5 Vulnerable settings in flash flood risk areas.

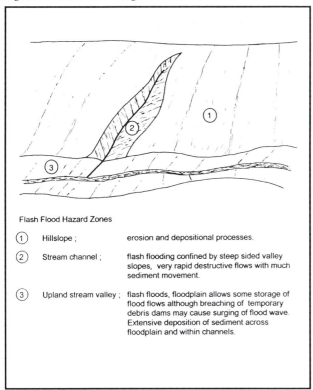

Flash Flood Hazard Zones

① Hillslope ; erosion and depositional processes.

② Stream channel ; flash flooding confined by steep sided valley slopes, very rapid destructive flows with much sediment movement.

③ Upland stream valley ; flash floods, floodplain allows some storage of flood flows although breaching of temporary debris dams may cause surging of flood wave. Extensive deposition of sediment across floodplain and within channels.

Table 5.5 lists a number of recent major flood events from the London area alone. Bearing in mind the cost of designing storm drains to even higher standards to take account of very intense short duration rainfall, it seems inevitable that some summer thunderstorms will lead to urban flood problems. However, short term retention of runoff from short, intense storms in areas such as minor roads, motorway hard–shoulders and car park can help avoid more difficult problems downstream of such paved areas.

Bridges and culverted sections can collect the debris swept downstream by the floodwaters, forming temporary dams. The subsequent build up of water and dam breach can cause destructive surges of water, greatly enhancing the destructive potential of a flood event. There is, therefore, a need for careful design and regular maintenance of trash screens and sediment traps upstream of culverted and other restricted stream sections.

The Significance for Conservation

Flash flood events are largely destructive by nature, although they may have a value in preparing gravel bed rivers for spawning fish (Institute of Hydrology, 1987). The resulting channel forms and debris accumulations can, however, be important

areas for geomorphological research, as reflected by the inclusion of a number of such sites in the Geological Conservation Review (Table 5.6). All these GCR sites may be vulnerable to the effects of river management aimed at reducing the risk from flash flooding.

Summary: The Significance for Planning and Development

Flash flooding is a significant problem for planning and development. Although events are most likely to occur in sparsely populated, steep catchments, the historical record clearly demonstrates that many communities situated close to streams or rivers around the margins of upland areas may be at risk. Such events can cause considerable damage to property and loss of life because of their sudden, unexpected nature, with peak flows often occurring within hours of the onset of heavy rain. In addition, flash floods can mobilise considerable amounts of sediment, initiate widespread channel erosion and lead to extensive deposition of material across floodplains downstream.

The occurrence of extreme climatic events, generally intense storms, is the main cause of flash flooding, although the potential effects of dam failures should never be underestimated. Indeed, the extreme nature of these events makes their prediction very difficult; frequently the estimated return periods are exceptionally large and their accuracy constrained by limited periods of rainfall or flow records. Major events are often regarded as "one-offs" or freak events, so there is no expectation of recurrence. As a result, although an area may be at risk from flooding it is often not perceived to be so. However, as was discussed in Chapter 2, a 1 in 1000 year event has a 3% probability of occurring within any ten year period (see Table 2.1).

Bearing in mind the major losses that have occurred in flash flood events, (as illustrated by the 1952 Lynmouth floods), **avoidance of vulnerable settings** is an obvious and potentially very effective planning response. Fortunately the areas that could be affected by floods of different severity are relatively narrow strips close to streams and can be readily defined from surface morphology adjacent to stream channels (Figure 5.5). Vulnerable settings could then be left undeveloped or utilised for recreational or leisure activities. In this context, it is important that planners and developers should be aware of the degree of risks associated with any

Table 5.5 Examples of recent flooding in the London area.

Date	Area	Comment
5 July 1985	Wimbledon, Westminster	20 minute downpour flooded the Wimbledon tournament on men's semi-final day, with water up to 1m deep. Elsewhere parts of the Palace of Westminster and County Hall were flooded, including underground stations at Swiss Cottage, Hyde Park and Loughton.
6 August 1981	Heathrow	Up to 1.5m of water flooded the airport; leading to power failures and disruption to air traffic.
	London	Kings Cross underground station was closed; passengers had to "paddle out of the station. Many roads flooded.
	Guildford	A3 trunk road closed because of 1m deep floodwaters.
30 May 1979	London	Traffic halted by thunderstorm, motorists trapped in floods. Services on many underground lines were suspended. North Circular road flooded.
14 August 1975	London	Serious flooding disrupted homes, businesses, road and rail communications. St Pancras railway station closed and services disrupted for a week. Damage estimated at £1M.
4 September 1974	London	North Circular at Brent Cross under 0.5m of water and closed to traffic. Flooding under railway bridge at West Drayton caused massive traffic jams on the approach roads to Heathrow.
7 August 1970	London	West End flooded; Duchess Theatre, Garrick Theatre and Victoria Palace Theatre closed; roads and shops flooded, manhole covers blew off because of the water pressure. Waterloo station flooded.

Table 5.6 Examples of Geological Conservation Review (GCR) sites created by flash flood events.

- the River Lyn catchment; scene of the 1952 Lynmouth Floods;

- Shaw Beck Gill, Northumberland; important upland record of historic floods and valley floor development;

- Black Burn, Cumbria; mid-channel bars, channel cut offs and boulder spreads from historic floods;

- Old Hamstocks Gullies, Lothian; gullying and slope erosion of a scale rarely seen in Britain;

- Luibeg Burn, Grampian; boulder bed stream with a documented history of large scale sediment movement in extreme floods;

- Quoich Water, Grampian; effects of the 1829 flood on the channel form;

- Allt Mor, Highland; flash flood deposits.

dams or reservoirs upstream (ICE, 1978).

It is important to note that development can have a significant effect on the risk of flash flooding, especially through:

- inadequate storm water drainage;

- the build up of water behind culverts and bridges, especially where these structures are blocked by debris brought downstream in high flows.

It is important, therefore, that local planning authorities seek and take full account of advice from the relevant drainage authority to ensure that drainage systems for proposed developments are adequate and include, where appropriate, surface or subsurface storage measures to delay the discharge of water into natural watercourses or the storm drainage systems. In addition, planners should take account of recommendations for debris screens or sediment traps upstream of culverted sections or other restricted channel sections and make provision for access for their maintenance.

Flash floods have created valuable GCR sites, frequently associated with unique boulder spreads or in-channel bars; planners and developers should be aware of the importance of these sites and their sensitivity to the effects of development or flood defence works.

Chapter 5 : References

Acreman M.C. 1985. The effects of afforestation on the flood hydrology of the upper Ettrick valley. Scottish Forestry 39, 89–99.

Acreman M.C. 1988. Hydrological analysis of the Truro floods of January and October 1988. Hydrological Data UK Yearbook, 27–33.

Acreman M.C. 1989. Extreme rainfall in Calderdale, 19 May 1989. Weather 44, 438–445.

Acreman M.C. 1990. Flood frequency analysis for the 1988 Truro Floods. J IWEM 4, 62–69.

Bleasdale A. And Douglas C.K.M., 1952. Storm over Exmoor on August 15, 1952. Meteorological Magazine, 81, 353–367.

Bossman-Aggrey P., Green C.H. and Parker D.J. 1987. Dam safety management in the United Kingdom. Middlesex University Geography and Planning Paper No. 21.

Calder I.R. and Newson M.D. 1979. Land use and upland water resources in Britain – a strategic look. Water Resources Bull. 15, 1628–1639.

Charles J.A. 1992. Embankment dams and their foundations: safety evaluation for static loading. Keynote Paper International Workshop on Dam Safety Evaluation, Grindelwald Switzerland.

Cross D.A.E. 1967. The Great Till Floods of 1841. Weather, 22, 430–433.

Howe G.M., Slaymaker H.O. and Harding D.M. 1967. Some aspects of the flood hydrology of the upper catchments of the Wye and Severn. Transactions of the Institute of British Geographers. 41, 33–58.

Institution of Civil Engineers, 1978. Floods and reservoir safety. Thomas Telford.

Institute of Hydrology, 1987. A study of compensation flows in the UK. Report No. 99.

Kidson C., 1953. The Exmoor storm and the Lynmouth floods. Geography, 38, 1–9.

Marshall W.A.L., 1952. The Lynmouth floods. Weather, 7, 338–342.

McFarlane D. and Smith C.G. 1968. Remarkable rainfall in Oxford. Meteorological Magazine, 97, 235–246.

Nairne D. 1895. Memorable floods in the Highlands during the Nineteenth Century. Northern Counties Printing and Publishing.

Newson M.D. 1975. Flooding and flood hazard in the UK. Oxford University Press.

Newson M.D., 1978. Drainage basin characteristics, their selection, derivation and analysis for a flood study of the British Isles. Earth Surface Processes, 3, 277–293.

Newson M.D. 1989. Flood effectiveness in River Basins: Progress in Britain in a decade of drought. In K Bevan and P Carling (eds) Floods : Hydrological, Sedimentological and Geomorphological Implications, 151–170. Wiley.

Newson M.D. and Maklin M. 1990. The geomorphologically-effective flood and vertical instability in river channels – feedback mechanism in the flood series for gravel bed rivers. In M.R.White (ed) River floor hydraulics, 123–140. Wiley.

Robinson M. 1981. The effects of pre-afforestation drainage upon the streamflow and water quality of a small upland catchment. Institute of Hydrology, Wallingford.

Robinson M. and Newson M.D. 1986. Comparison of forest and moorland hydrology in an upland area with peat soils. International Peat Journal. 1, 49–68.

Taylor S. 1994. Chichester's floods: a natural disaster? Geographical Magazine May 1994, 43–45.

Timber Growers Association 1986. Afforestation and nature conservation interactions.

Vaughan R.A. **1980.** The floods in South Wales of late December 1979. BSc dissertation, University of Wales.

Chapter 5 : Suggested Reading

Institution of Civil Engineers, 1978. Floods and Reservoir Safety. Thomas Telford.
Lewin J. (ed) 1981. British Rivers. George Allen and Unwin.
Newson M.D. 1975. Flooding and flood hazard in the UK. Oxford Univ. Press.
Ward R.C. 1978. Floods : a geographical perspective. Macmillan Press.

6 Rivers: Bank erosion, Sedimentation and Channel instability

The Nature of the Problems

Although British rivers are not subject to the sudden changes in course that are characteristic of some of the world's major rivers, they are frequently affected by small–scale channel changes in channel position and size. A wide variety of types of channel change can occur (Table 6.1) involving modifications to the course (planform), the channel size (cross section) and the gradient (longitudinal form). Often changes in one aspect of form tends to result in changes in the others. For example, an increase in the depth of a channel is often associated with an increase in gradient.

Perhaps the most significant changes are those associated with the reduction in channel cross section due to sedimentation which can lead to significant channel maintenance to ensure river flood capacity. In some instances, a long term and expensive maintenance commitment is required to maintain the design standard of flood defence capital works. For example, 5000–8000 tonnes of gravel have to be removed at least once every year from a flood defence scheme on the River Usk at Brecon, at an annual cost of £10,000. Sear and Newson (1992) have estimated that an annual total of £7M, (some 15% of the total NRA Main River maintenance budget) is spent on sediment–related problems in England and Wales. The problems are typically associated with 3 settings:

- low–cost, low to moderate frequency maintenance of **gravel and cobble bed channels** around the margins of upland areas. Many gravel–bed rivers are managed by routine maintenance in areas of shoals or eroding bends. This generates large numbers of small maintenance commitments;

- high-cost, low frequency maintenance on **silt and clay channels** in lowland areas.

Siltation is progressive and maintenance operations are implemented as rolling programmes covering long reaches of an individual channel.

- high cost, high frequency maintenance of **tidal channels**, where channel banks are developed in weak materials.

The scale of channel change ranges from a few centimetres to around 10m a year. Only in certain settings can channel migration processes be considered large enough to be **significant** within the context of this study. In practice, the threshold tends to be whether changes can be identified from available maps; one channel width per 100 years is considered to be lower limit of significant instability.

Hooke and Redmond (1989) found that of 120 mainly upland streams, 41 show a major "natural" change. At least 800km of river in England and Wales have been affected by major change in course, with notable examples including:

- a net increase in width of the braided River Feshie of 7m over five years, (Ferguson and Werrity, 1983);

- 3m vertical incision of the River Ystwyth since 1930, (Lewin et al., 1983);

- 4m lateral erosion of the River Dane in Cheshire in 1986, immediately downstream of a natural cutoff formed in 1984 (Hooke and Redmond, 1989).

Although river channel changes are not perceived to be a major hazard in Great Britain, they can result in a range of locally significant problems. An obvious impact is the loss of land that results from bank erosion, meander migration and braiding. In most cases the loss of land is an agricultural problem, with the actual costs incurred depending

Table 6.1 Types of channel change.

Planform

Confined or restricted migration involves downstream translation of meandering channel with little net change in form characteristics in a reach. Movement is often constrained by terraces or valley wall. In low rates of migration erosion occurs only on bend apexes.

Active meandering may involve a net change in form, either an increase or decrease in channel length in a reach. Major components of change in meanders are: migration, extension, compound forms and cutoffs. Meanders may undergo an evolutionary sequence from migration of simple loops through to cutoff of compound loops.

Meander-braiding switching. Channels may change from single to multiple and vice versa. The threshold is influenced by discharge, gradient and sediment load. Floods may cause braiding where sediment load is increased. Channels may cease to be braided when sediment supply is decreased e.g. after period of mining.

Braiding. Degree of braiding, indicated by number of bars and width of channel zone may alter. Braided reaches tend to be unstable in form and are influenced by individual floods. Frequent channel switching takes place by avulsion and deposition.

Cross-Sectional Form

Width increase takes place by erosion of both banks, e.g. in response to urbanisation. Decrease takes place by bar deposition and formation of berms e.g. downstream of some reservoirs.

Depth increase results in incision of channel and can in turn destabilise channel banks. Incision may take place in large floods and if bed armouring is removed. Decrease results from net deposition from increased sediment supply or decreased gradient.

Longitudinal Form

Steepening results from incision and increase in slope upstream. A sharp break of slope (headcut or knickpoint) may be created and this may progress upstream over time.

Flattening of slope results from aggradation or build-up of sediment controlled either from upstream supply or a raised local base-level downstream (e.g. sea level rise).

Channel change takes place in three-dimensions and therefore instability involves combinations of these type of changes.

on the land use and value. However, in the case of the lateral shift of a channel, erosion will usually be balanced by deposition. This may not always be to the benefit of the affected landowner where the river channel acts as a property boundary. In addition, it may take several years before the deposited area becomes as productive as the lost land.

Channel instability can also lead to difficulties at stream or river crossing points. Bridges and road and railway embankments are probably the most vulnerable structures, especially during flood events. Examples include the loss of a section of railway embankment at Dalguise and structural damage to Fortevoit rail bridge during the 1993 Perth floods; in October 1987 four people died when channel scour and erosion around the piers of a railway bridge at Glanrhyd, Dyfed led to the bridge collapsing under the weight of a train.

Riverside property can be damaged by bank erosion but, fortunately examples are rare; during the July 1958 floods in Sheffield, South Yorkshire a terraced house collapsed into the River Sheaf, undermined by the flow.

Breaching of floodbanks can occur in major floods and can cause the formation of new channels. The 1990 and 1993 flood on the River Tay, for example, caused significant changes, including the creation of small sections of new channel across the floodplain (Gilvear and Harrison, 1991; Gilvear et al, 1994). Five metres of erosion occurred in the 1990 flood on the outside of the meander bend at Braecock over a 300m reach during the flood. In addition, two discrete breaches occurred in the river embankment causing over 12km² of farmland to be flooded. The initial breaching of the flood embankments resulted from scouring of gravel from beneath the flood embankment foundations.

Severe erosion and cutting of channels into the floodplain deposits, up to 2m deep, were associated with the floodwater movement. The north–south trending erosional channel severed the access road to Braecock Farm (Figure 6.1). Both breaches correspond with former river channel courses, a pattern that was repeated on a much larger scale during the 1993 floods when many of the embankment failures took place where the present course crossed abandoned channels.

It is clear from these examples that the more severe problems tend to be associated with flood events. However, erosion and deposition is not restricted to extreme events. It can be a continuous process, undermining bankside structures, and causing scour around bridge piers. Few of these impacts are dramatic and often go unreported; it is, therefore, difficult to quantify the scale of the losses associated with these processes. An indication can be given, however, from a consideration of the works that may be needed to protect vulnerable structures. Hemphill and Bramley (1989) report from a survey of authorities that there are over 2400km of bank protection in the UK, equivalent to about 3% of the total length of water courses protected on both sides.

River channel changes and bank erosion can promote slope instability on neighbouring slopes, as has occurred along much of the lower Tees valley between Darlington and Thornaby. Fortunately many of the resulting landslide features are in rural areas where they have not had significant consequences. Problems have however arisen in Yarm where continuing erosion has initiated small landslides on the low river cliffs developed in glacial materials and led to serious damage to a number of properties (Geomorphological Services Ltd, 1989).

Instability caused by river erosion has not, however, been a widespread problem to development in Great Britain. One of the few examples described by Jones and Lee (1994) is the Cliff landslide on the banks of the River Irwell in Salford has caused recurrent damage to the road and property immediately upslope over the last 100 years (Harrison and Petch, 1985; Figure 6.2). The first reported movements occurred in February 1882 when a large slide was recorded by the City Engineers Department. By May of the same year, cracks had appeared in the road immediately upslope of the slide, and subsequently a drainage scheme was installed. Continued ground movement problems resulted in the tram service along the road above the slide being discontinued in 1925,

and a section of the road closed to motor traffic in 1926. Further movements have occurred intermittently and have caused damage to a row of terraced houses at the head of the slide.

From this brief introduction, it is clear that the impacts of channel changes and bank erosion are generally localised, affecting property, infrastructure and services at a vulnerable point along an unstable river section. However, downstream problems can arise most notably as a result of:

●   mobilisation of river channel or bank sediments which can be deposited further downstream and giving rise to flood risk or navigability problems elsewhere;

●   channel changes at one point can lead to changes downstream as the channel attempts to adjust to the resulting changes in width, depth, gradient or sediment concentration. This is a particularly important consideration when constructing bank protection measures which may simply pass the problems downstream as the channel attempts to adjust to these artificial constraints (Figure 2.21).

The Causes of Sedimentation and Channel Instability

The main factors influencing the occurrence of sedimentation and channel instability are:

●   **site conditions**
●   **catchment character**
●   **climate**
●   **land management**

On any stretch of river, erosion and sediment transport will occur where the force of the water flow exceeds the resistance of the river bed or bank, i.e. a function of the **erosivity** of the running water and the **erodibility** of the materials. Thus, channel instability and bank erosion tends to occur along reaches developed in soft alluvial or superficial deposits (Chapter 2). The morphology of such **alluvial rivers** depends on the river regime and sediment yield; whilst these remain broadly constant in a catchment, the gross river pattern may remain stable even though the channel position itself may not be static (Richards, 1982). River channel form is adjusted to the catchment processes, and is therefore said to be in "**dynamic equilibrium**".

Figure 6.1 The channel changes associated with the 1990 Tay floods (after Gilvear and Harrison, 1991).

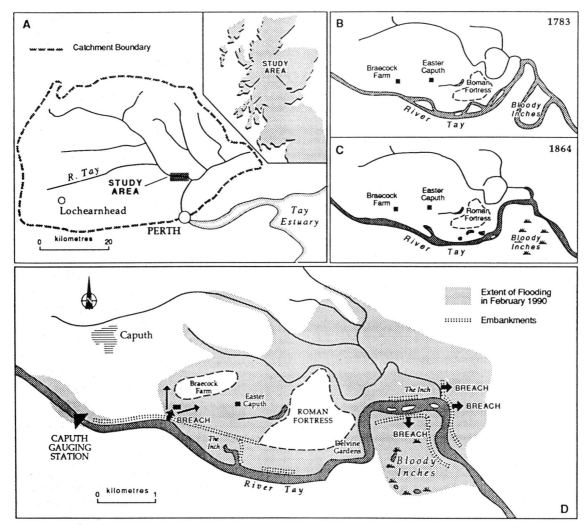

Channel morphology will adjust to changes in water and sediment discharges associated with:

(i)   **long term influences** including environmental change, which cause gradual, progressive adjustment of the channel to maintain dynamic equilibrium. In general, high rainfall periods such as occurred at the end of the "Little Ice Age" are believed to be associated with increased erosion, lateral instability, incision and channel widening. Macklin et al (1991), for example, have noted profound channel changes in northern England during this period. However, the effects of environmental change may be delayed; a number of rivers such as the South Tyne are currently incising into their flood plains in what is believed to be a continuing response to the climatic shift since the "Little Ice Age".

(ii) **medium term adjustments**, frequently associated with human activities causing a temporary disequilibrium in the channel which then passes through a series of transient states in reaching a new equilibrium form. The responses are, however, often complex and dependent on site conditions. Of particular importance is the influence of:

●   **land use change**; through the increase in runoff and hillslope erosion. However, it is likely that much of the sediment derived from increased slope erosion is "stored" at the base of the slope, with only a small proportion reaching the channel. The significance of historic or recent land use change is by no means clear with little direct evidence available to evaluate its role in promoting periods of instability.

70

Figure 6.2 The cliff landslide, Salford (after Harrison and Petch, 1985).

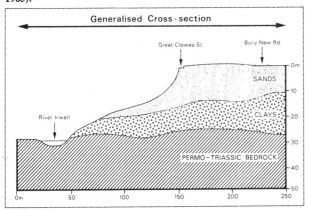

reservoirs and dams; significant changes in both discharge regime and sediment load take place as a result of water impoundment by reservoirs. Changes of this type in Britain have been extensively investigated, (e.g. Petts 1979, 1984). Generally dams decrease peak flows downstream and impound practically all sediment (see Chapter 3). Channels tend to reduce in size downstream but also tend to incise because of the lack of sediment. However, it has been found that the response may be much more complex than this especially progressing downstream where the main channel interacts with tributaries. At stream confluences, sediment inputs from unregulated watercourses can build up in the main channel where the flood peak is reduced. Dredging and bank protection is then needed to maintain capacity, as has occurred on the North Tyne and River Dee.

Most of the adjustment seems to be by cross–section form and slope but even along one river there may be both increases and decreases in channel capacity. Channel degradation and scour will persist until the reduction of channel slope reduces the flow velocity below the threshold for sediment transport. The timescales of adjustment to river impoundment also vary but Petts (1984) has stated that five years may be required before any channel response is observed;

continuing channel changes have been reported more than 50 years after dam construction, and stability in terms of sediment transport and channel form may take 200 years.

- **channelisation**; changes have arisen as a direct result of river works, in some cases exacerbating the problems they were meant to solve. The impact depends on the nature of the works and the sensitivity of the river, but frequently involves change in cross section or gradient due to increased erosion.

- **waste disposal**; former mine workings in upland areas frequently disposed of spoil and waste in mounds adjacent to river channels or directly into the stream itself. The resulting increase in sediment load can have an impact on its stability as shown on the Tyne and in Welsh valleys (Lewin et al 1983, Lewin and Macklin 1987, Macklin and Lewin 1989). Generally, increased sediment will lead to increased instability, tendency for braiding, aggradation and channel switching. Sediments from metal mines may be toxic enough to decrease vegetation growth and thus contribute to instability. The reduction in sediment loads following mine closure can reverse some of the channel changes. Lewin et al (1988) have shown that cessation of mining is still causing adjustment of the channel 50 years later.

- **mineral workings**; removal of gravel from the channel on the Tywi has resulted in a wider braided channel and is reported to have loosened bed sediments which could be more easily eroded, leading to sedimentation and channel widening downstream (Lewin et al 1988). The effects can be transmitted over several kilometres and may take over 50 years to occur.

71

- **other causes**; a range of other causes have been cited as causing channel instability or bank erosion. These include the effects of boats on navigable rivers, construction of weirs, water abstraction, e.g. for water supply, and the effects of bridges. Water abstraction leads to alteration of discharge and can therefore induce adjustments of the channel. Wiers and bridges tend to have localised effects, the former tending to stabilise channels but inducing instability where they collapse, e.g. weirs associated with old mills. Boats may cause some bank erosion but not generally large-scale instability (Hooke et al 1991).

(iii) **short term responses** may follow individual extreme flood events, as described earlier for the Tay floods of 1990 (see also Figure 6.3 for changes resulting from the 1982 floods in The Howgill Fells; Harvey, 1987).

The channel responses will be complex and very difficult to predict. Schumm (1969), however, has presented a qualitative approach which indicates the direction in which key channel variables are likely to change (Figure 6.4). Thus, a rise in runoff and discharge following urbanisation could lead to increases in width, depth and, hence, channel capacity. An increase in sediment supply, as occurred during heavy metal mining activity in mid-Wales when spoil was deposited by stream channels, can lead to an increase in channel width and gradient, but a decrease in depth; this period of mining activity was also accompanied by floodplain deposition (Lewin et al 1988). It is important to note, however, that there are many instances where an activity in a catchment or channel reach has not produced a measurable effect.

The Identification of Vulnerable Areas

The distribution of potential sediment transport problem sites in England and Wales, has been identified by Sear and Newson (1992) who conducted a survey of NRA Regional staff. Problems are typically associated with gravel bed channels, silty clay channels in lowland areas and tidal channels (Figure 6.5); particular problems

may occur on meandering or straightened channels, and at river confluences within upland areas. The distribution of recorded areas of significant bank erosion and channel instability is shown on the accompanying 1:625,000 scale thematic maps. This map is based on a number of sources:

- published scientific papers and academic research projects (e.g. Hooke and Redmond, 1989; Brookes and Gregory 1983; Lewin, 1983);

- discussions with local authority engineers.

The pattern of unstable channels highlights a concentration around the margins of upland Britain, most notably: Northern Britain, the Welsh borders and South West England (Table 6.2). In these areas instability tends to occur in the middle reaches of a river, where it emerges from the uplands. Here rivers are still sufficiently steep to have potential for erosion, whilst entering alluvial areas where the banks and bed are readily erodible.

Within these regions it is likely that most rivers will have unstable reaches, although there will be considerable variation in the scale and nature of the instability. The Severn, for example, shows a range of channel patterns and instability features in its upper reaches (Figure 6.6), including:

Table 6.2 Examples of notable unstable rivers.

Zones of major lateral instability, where rates of 0.5–1m/year and large scale change in channel location have been recognised occur in the middle reaches a number of rivers, including:
(i)　　　　Northern Britain, for example:
- the Feshie
- the Spey
- the Tay
- the South Tyne
- the Till
(ii)　　　the Welsh borders, for example:
- the Severn
- the Wye
- the Teme
- the Dane
- the Dee
(ii)　　　South West England, for example:
- the Exe
- the Culm
- the Axe

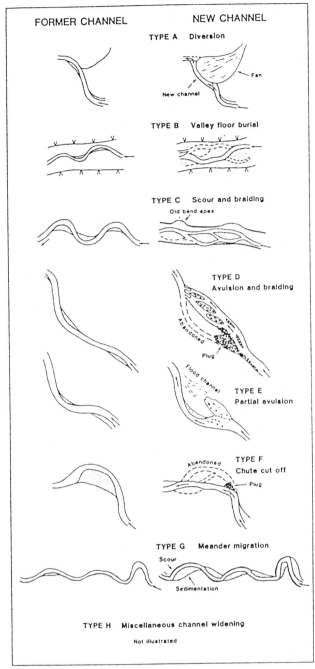

FORMER CHANNEL   NEW CHANNEL

TYPE A   Diversion

TYPE B   Valley floor burial

TYPE C   Scour and braiding

TYPE D   Avulsion and braiding

TYPE E   Partial avulsion

TYPE F   Chute cut off

TYPE G   Meander migration

TYPE H   Miscellaneous channel widening
Not illustrated

(i)   at Penstrowed, a high-sinuosity channel has been replaced, over 70 years, by a straighter section;

(ii)   at Llandinam, there has been downstream movement of in-channel gravel bars with minor phases of erosion and sedimentation;

(iii)   at Maesmawr, channel migration has involved down-valley shift, lateral expansion and chute cutoff;

(iv)   at Welshpool, a sinuous channel in 1775 was replaced by a straighter one involving

braiding, probably due to an extreme flood event in 1795. During the 20th century the series of meanders have developed, partly constrained by the construction of Leighton Bridge in 1871.

Other sections of the Severn show no signs of natural change, being either inactive or modified as a result of channelisation. For example, the lower part of the Trannon, a 72km² left-bank tributary of the Severn, has been converted from a rather stable low-gradient meandering channel in the early nineteenth century (Figure 6.7) to one in which straight and sinuous segments alternate, and where erosion is now a problem (Levin, 1987). Part of the earlier 'natural' system appears to have been a multithread pattern. The channelisation schemes were initially associated with railway development, designed to alleviate a flood problem.

Within unstable reaches there are often marked zones of sedimentation and instability, separated by more stable areas. On the Tyne, for example, stable sections occur where the river crosses bedrock and, hence, is more confined. Similarly on the River Dee in Scotland instability occurs in less confined reaches between rock gorges (McEwan 1989) and sedimentation zones are recognised by Gilvear and Winterbottom (1992) on the Tay. These unstable reaches are usually of the order of one to several kilometres in length, e.g. on the River Tywi alternating lengths of 1–1.5km stable and unstable channel occur (Smith 1989). As adjustments to channel form tend to occur over short sections, it is difficult to identify particularly vulnerable settings within an unstable reach. Figure 6.8 clearly demonstrates the very localised nature of channel movement, with considerable variability even over short distances.

Fewer unstable channels have been identified in the east Pennines, the north west Pennines, the Lake District and the North York Moors. Although examples do occur, they tend to involve short isolated unstable reaches separated by long stable sections, as on the Rivers Ure and Swale in Yorkshire.

Channel change directly attributed to human interference is very localised, affecting the cross-sectional form and slope rather than planform. The greatest adjustments tend to be on those streams in areas with a higher liability to instability. Within catchments, channel instability caused directly by human impact often does not have a simple downstream distribution from the location of the disturbance. Indeed, effects can also be propagated

73

Figure 6.4 Likely channel changes associated with changes in discharge and sediment load (after Schumm, 1969).

	Channel width	Channel depth	Channel gradient	Meander wavelength	Channel sinuousity
Increased discharge	●	●	○	●	
Increased sediment load	●	○	●	●	○
Increased discharge and sediment load	●	□	□	●	○
Decreased discharge	○	○	●	○	
Decreased sediment load	○	●	○	○	●
Decreased discharge and sediment load	○	□	□	○	●

Key:

   ●   Increase

   ○   Decrease

   □   Increase or decrease

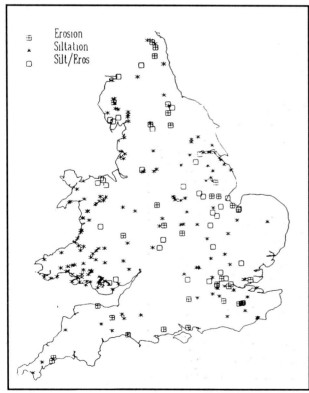

Figure 6.5 The distribution of river channel problem sites in England and Wales (after Sear and Newson, 1992).

upstream. Of reaches adjusting subsequently to upstream channelisation, the extent of the downstream effects ranged from 120m to 1952m. There is no well defined decline in the magnitude of effects with distance downstream (Brookes 1987). Similarly the effects of reservoirs have been recorded for between 250m and 10km downstream of the dam, but the spatial distribution of response can be very complex.

The Effects of Development

It is clear that river channels can be very sensitive to human activity, through:

- **direct** effects caused by deliberate river management to control streamflow, reduce flood risk and improve navigability;

- **indirect** effects resulting from land use changes which lead to alterations in sediment yield and runoff.

Figure 6.6 Channel changes on selected reaches of the Upper Severn (after Lewin, 1987).

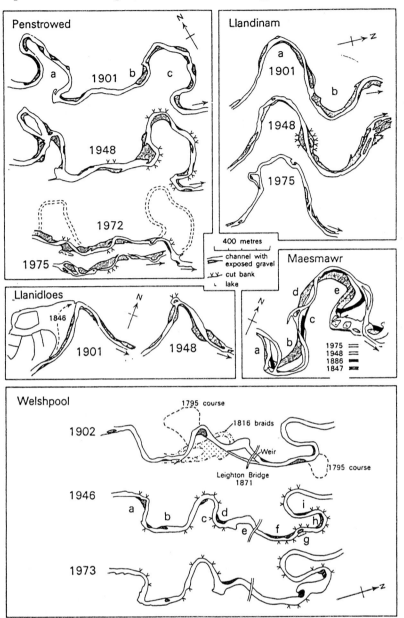

Figure 6.7 Present and former courses of the Lower Trannon (after Lewin, 1987).

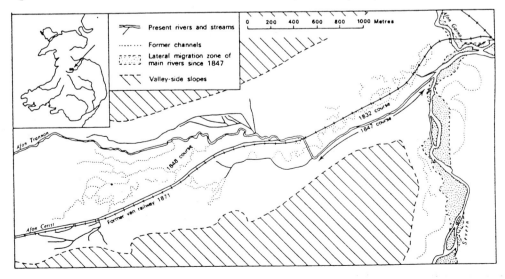

Figure 6.8 The pattern of channel change on the River Culm and River Axe, Devon (after Hooke, 1977).

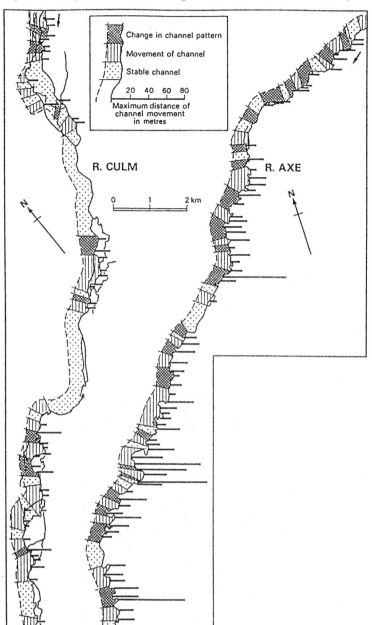

It is widely recognised that urbanisation can lead to channel modifications downstream following an increase in the size of peak flows and a decrease in sediment supply (e.g. Knight 1979; Hollis and Luckett 1976). Amongst the most common changes are bed and bank erosion, loss of riverbank trees and undermining of structures. In a review of channel changes following the development of Cumbernauld New Town in Strathclyde, Roberts (1989) noted that some streams had enlarged through the urban area. Locally there were instances of extensive erosion and bank collapse; up to 10m of vertical incision through glacial till was observed in gulleys draining industrial areas. Some channels developed gravel bars and became more braided.

River channel engineering has had notable upstream and downstream effects throughout Great Britain; Figure 6.9, for example, shows the extent of 8500km of capital works and major improvement schemes, many of which may be regarded as having a major and lasting impact on channel morphology (e.g. embankments, bank protection, channel enlargement or realignment, culverting etc; Brookes and Gregory 1988). A further 35,500km is maintained which may involve dredging of channel sediments and vegetation control. Such **channelisation** generally involves attempting to control the channel form at a site, it can initiate instability not only in the improved channel reach but also upstream and/or downstream as the channel adjusts to a new state of equilibrium. Indeed Brookes (1985, 1987)

Figure 6.9 River channelization in England and Wales (after Brookes et al, 1993).

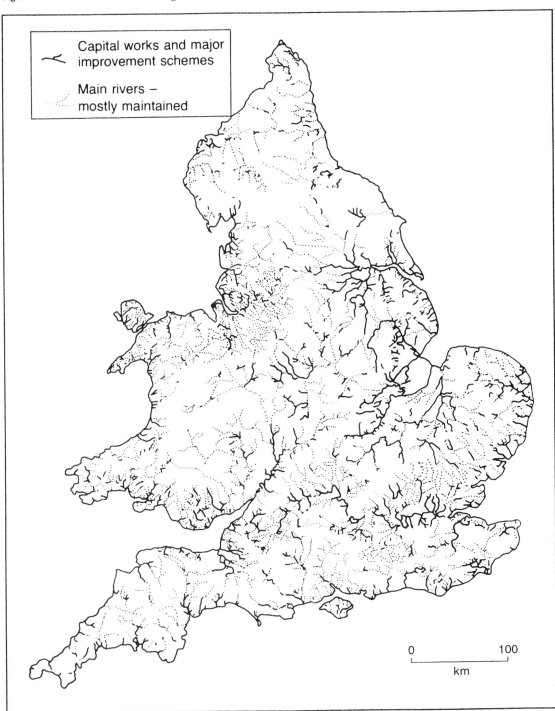

identified that many channelised sites in upland Britain have downstream reaches with clear indications of erosion and that channels had widened over time at these sites.

The Significance for Conservation

River beds, banks and floodplains are all important in supporting particular species or communities of plants and animals throughout or during parts of

their life cycle. For example, some fish require shallow water and gravels, as found on riffles, in which to spawn but commonly live in the deeper water at other times. Amongst the most important features of the river environment are:

(i)   the flood pulse in determining many ecological functions both within the floodplain itself and the associated ecosystems;

(ii)   topography and site characteristics;

77

(iii)   water quality;

(iv)   the role of geomorphological dynamics in determining the mosaic of ecological patches that make up natural river corridors.

Channel changes can be vital to the maintenance of both channel and floodplain sites. Floods tend to maintain ecological diversity, by removing vegetation which may have clogged a channel, by recreating bare vertical banks which have become overgrown, or by forming new riffles and pools (Lewis and Williams 1984). In addition, maintenance of natural vegetation and ecological diversity can help maintain water quality through natural processes of purification. In this context, various sources of guidance are available for river managers on fluvial habitats and the species found in each type, with the key characteristics of the sites identified (e.g. Lewis and Williams 1984, Newbold et al 1983, Boon et al 1992).

The nature and value of many rivers is illustrated by Newbold et al (1983) who note that:

"In erosive sections of rivers, vertical banks of 3m or more facing shingle beds on the opposite side frequently occur. These vertical banks are colonised by mosses and numerous annual and deep-rooted perennial flowering plants. They also provide nesting sites for kingfishers and sand martins which dig horizontal tunnels in the soft sediment. The shingle banks often develop rich, but temporary, floras which create a profusion of spring and summer colour ... Shingle banks are used by the common sandpiper and oystercatcher as breeding or feeding grounds. The darting wagtails feed on airborne insects whilst the dipper patrols underwater and flycatchers often use the overhanging branches as lookout posts. The otter is found in such habitats." (Newbold et al, 1983).

The key habitats characteristic of unstable river channels are therefore, weed-free gravel beds, vertical bare banks, sandy and gravelly bars without luxuriant vegetation, and old cutoffs at various stages of infill from extensive water areas to marsh and slightly wetter zones of the floodplain. Meander cutoffs will tend to silt up over time if natural flooding processes are allowed to continue. This is inevitable and may be regarded by some as loss of a rich habitat. However, in such a situation it is likely that the channel instability will continue to create new cutoffs and thus the range and diversity of habitat will be maintained and enhanced. The conservation need is, therefore,

for the continuance of the processes not for preservation of forms created.

Many riverine biological SSSIs have been designated in Britain. The criteria used in selection were: naturalness, typicality or representativeness, diversity, rarity, size or extent of site, fragility, and geographical position. Unstable river reaches score highly on some of these criteria, particularly naturalness and diversity. In the guidance on management of river SSSIs Newbold et al (1983) note that it is important to retain riffle/pool sections, sandy meanders and braided channels; these are characteristics of unstable river zones.

Channelisation of mobile river reaches commonly involves direct destruction and removal of plants and habitats, particularly the loss of pools and riffles, vertical eroding banks and sinuous courses. The impact on fish is highlighted by Brookes (1988). Changes in fish populations can result from the loss of natural rifle-pool sequences which provide a variety of low flow conditions suitable as cover for both fish and the organisms on which they feed Fish require sheltered water in fast flowing rivers, these conditions may be absent where a meandering stream has been artificially straightened. Changes to the channel width and depth may create unsuitable habitats or present topographical difficulties for fish migration. Clearance of bankside vegetation may destroy valuable cover for fish.

Unstable river sections can also be of conservation interest for their geomorphological value. Indeed, the Geological Conservation Review has identified a range of river corridor sites noted for the features associated with active channel migration including:

●   R. West Allen, Tynedale;
●   Alston Shingles, R. South Tyne;
●   R. Nent, Blagill;
●   Langden Brook, Lancashire;
●   R. Dee at Holt – Worthenbury.

It is worth noting that the designation of unstable channel reaches for their conservation value can lead to conflict with flood defence and landowning interests. For example, a part of the then NCC's Geological Conservation Review, the stretch of The River Dee between Holt and Worthenbury has been identified as an area of particular geomorphological interest. The Dee is one of only a few large upland rivers with a well developed, mobile meander belt in its lower course that is relatively free from direct human intervention. The reach includes a range of meander forms with numerous cut-offs and abandoned channels. The Countryside Council for Wales (CCW; the successor to the NCC in Wales) has proposed to recommend to the Secretary of State that the stretch be designated as an SSSI.

Average bank and erosion rates along sections of the river have been less than 1m/year, although some meanders may be more mobile in the short term. Analysis of the historical river channel planform changes has indicated that the channel has appeared to become more stable in recent years, probably due to upstream regulation works. River channel migration and regular flooding have been seen as a problem by the local farming community. As part of their flood defence responsibilities the NRA and its predecessors have constructed bank stabilisation works including toe reinforcement using boulders, stone pitching across the bank faces and, more recently regrading and strengthening using geotextiles. However, concerns have been expressed that the measures are likely to increase rather than reduce channel change and conflict with the conservation objectives of CCW.

As part of the SSSI notification procedure CCW must specify those operations likely to damage the special interest. Once notified, owners and occupiers must give CCW written notice before carrying out a potentially damaging operation (PDO). On the River Dee the procedure for notification has been delayed because of a disagreement between the NRA Flood Defence Committee and landowners and CCW over the most appropriate management strategy for the river that takes into account the conservation interests as well as the protection of agricultural land.

Summary: the Significance for Planning and Development

Bank erosion, sedimentation and river channel instability is a feature of many British rivers, especially those flowing through soft superficial deposits and alluvium around the margins of upland areas. In general, the problems that can arise are localised, frequently associated with loss of land, undermined structures and scouring around bridges. This can, of course, have serious consequences, as occurred in October 1987 when an underscoured railway bridge at Glanrhyd, Dyfed collapsed under the weight of a train and 4 people drowned.

River erosion can initiate land instability an adjacent valley slopes, as at Yarm on the lower

Tees and at the Cliff in Salford. These are, however, amongst the few reported significant incidents of urban landslides caused by river erosion in Britain; the majority of such failures are in rural areas where the consequences are often minimal. Erosion can also lead to the mobilisation of sediments from within the channel or from the banks; this material can be deposited downstream and, thus, contribute to flood defence maintenance and navigability problems elsewhere.

The causes of channel changes are very complex, ranging from long term progressive adjustments to environmental change to short term responses to extreme events, as occurred on the Lower Tay in 1990 (Figure 6.1). Man can have an important influence on channel changes through:

- **direct effects** caused by deliberate river management or channelisation works;

- **indirect effects** resulting from land use change within a catchment.

Urbanisation can have notable consequences for river channel instability both at the immediate site and elsewhere (both upstream and downstream), following the accompanying increases in runoff and decline in sediment supply. Channelisation frequently leads to important channel changes as artificial modifications to channel morphology tend to cause adjacent sections to adjust to the resulting changes in flow and sediment load. Indeed, river engineering works to protect against erosion at one point will almost invariably result in an increase in the rate of bank erosion elsewhere.

It is clear that channel changes can present constraints to land use and development in certain areas. The planning system can be used to minimise the impacts resulting from these processes by ensuring that particularly unstable channel sections are avoided by development. This may involve assessing the likely trends for future channel behaviour and defining "set-back" zones within which there could be a significant risk of undermining during the lifetime of a structure. The local planning authority should seek advice from the relevant drainage authority when any channel works are to be considered.

In other areas, local planning authorities should ensure that riverside developments are adequately protected from the impact of bank erosion and channel instability. However, the potential effects of any preventative measures on upstream or downstream channel sections would need to be considered when determining planning applications, most notably the possibility of increasing erosion or sedimentation problems elsewhere. In addition, planners should bear in mind the important role that "natural" channel instability and bank erosion plays in creating and maintaining habitats and geological features of national importance. These valued features of the river environment can be very sensitive to the channel changes that may follow channelisation works. In this context, planners should always liaise with drainage operators and riparian owners to ensure the conservation interest of river sections is not adversely affected by inappropriate channelisation works.

Chapter 6: References

Boon P.J., Calow P. and Petts G.E., 1992. River Conservation and Management. John Wiley and Sons.

Brookes A.B. 1985. Downstream morphological consequences of river channelisation in England and Wales. Geographical Journal, 151, 57–62.

Brookes A.B. 1987. River channel adjustments downstream from channelisation works in England and Wales. Earth Surface Processes and Landforms, 12, 337–351.

Brookes A.B. 1988. Channelized rivers : perspective for environmental management. Wiley and Sons.

Brookes A., Gregory K.J. and Dawson F.H. 1983. An assessment of river channelization in England and Wales. The Science of the Total Environment, 27, 97–111.

Brookes A. and Gregory K.J. 1988. Channelization, river engineering and geomorphology. In J.M. Hooke (ed.), Geomorphology in Environmental Planning, (145–168, Wiley).

Geomorphological Services Ltd 1989. Tees Barrage and Crossing Bill: Geomorphological Survey report. Confidential Report.

Gilvear D.J. and Winterbottom S.J. 1992. Channel change and flood events since 1783 on the regulated River Tay, Scotland: Implications for Flood Hazard Management. Regulated Rivers: Research and Management, 7, 247–260.

Gilvear D.J. and Harrison D.J. 1991. Channel change and the significance of floodplain stratigraphy : 1990 flood event, Lower River Tay, Scotland. Earth Surface Processes and Landforms. 16, 753–762.

Gilvear D.J., Davies J.R. and Winterbottom S.J., 1994. Mechanisms of floodbank failure during large flood events on the rivers Tay and Earn, Scotland. Quarterly Journal of Engineering Geology, 27, 319–332.

Harrison C. and Petch J.R. 1985. Ground movements in parts of Salford and Bury, Greater Manchester – aspects of urban geomorphology. In R.H. Johnson (ed) The geomorphology of North-West England 353–371, Manchester Univ. Press.

Harvey A.M. 1991. The Influence of Sediment Supply on the Channel Morphology of Upland Streams: Howgill Fells, Northwest England. Earth Surface Processes and Landforms, 16(7), 675–684.

Harvey A.M. 1987. Sediment Supply to Upland Streams: Influence on Channel Adjustment. In C.R. Thorne, J.C. Bathurst and R.D. Hey (eds), Sediment Transport in Gravel–bed Rivers, 121–150. Wiley & Sons.

Hemphill R.W. and Bramley M.E. 1989. Protection of River and Canal Banks – a guide to selection and design. CIRIA Water Engineering Report. London, Butterworths.

Hollis G.E. and Luckett J.K. 1976. The response of natural streams to urbanisation: Two cast studies from south east England. Journal of Hydrology 30, 351–363.

Hooke J.M. 1977. The destruction and nature of changes in river channel patterns. In K.J. Gregory (ed.), River Channel Changes, 265–280. Wiley.

Hooke J.M., Bayliss D.H. and Clifford N.J. 1991. Bank Erosion on Navigable Waterways. Report to National Rivers Authority. University of Portsmouth.

Hooke J.M. and Redmound C.E. 1989. River channel changes in England and Wales. Journal of Institution of Water and Environmental Management, 3, 328–335.

Hooke J.M. and Redmound C.E. 1992. Causes and nature of river planform change. In P. Billi, R.D. Hey, C.R. Thorne and P. Tacconi (eds), Dynamics of Gravel–Bed Rivers, 549–563. Chichester: Wiley.

Jones D.K.C. and Lee E.M. 1994. Landsliding in Great Britain: a review. HMSO.

Knight C.R. 1979. Urbanisation and natural stream channel morphology: the case of two English new towns. In G.E. Hollis (ed) Mans impact on the hydrological cycle in the United Kingdom. Geobooks, 181–198.

Lewin J. 1983. Changes of channel patterns and floodplains. In K.J. Gregory (ed), Background to Palaeohydrology, 303–319. Wiley & Sons.

Lewin J., 1987. Historical river channel changes. In K.J. Gregory, J. Lewin and J.B. Thornes (eds.). Palaehydrology in Practice – A River Basin Analysis, 161–176, Wiley and Sons.

Lewin J. and Macklin M.G. 1987. Metal Mining and Floodplain Sedimentation in Britain.

Lewin J. and Macklin M.G. 1987. Metal Mining and Floodplain Sedimentation in Britain. In V. Gardiner (ed), International Geomorphology, 1009–1027. Wiley & Sons Ltd.

Lewin J., Bradley S.B. and Macklin M.G. 1983. Historical valley alluviation in mid–Wales. Geological Journal, 18, 331–350.

Lewin J. Macklin M.G. and Newson M.D. 1988. Regime Theory and Environmental Change – Irreconcilable Concepts? In W.R. White (ed), International Conference on River Regime. Wiley & Sons.

Lewis G. and Williams G. (eds) 1984. Rivers and Wildlife Handbook. Lincoln: RPSB.

McEwen L.J. 1989. River Channel Changes in Response to Flooding in the Upper River Dee Catchment, Aberdeenshire, over the Last 200 Years. In K. Beven and P. Carling (eds), Floods, 219–238. Wiley & Sons.

Macklin M.G. and Lewin J. 1989. Sediment transfer and transformation of an alluvial valley floor: the River South Tyne, Northumbria, UK. Earth Science Processes and Landforms, 14, 233–246.

Macklin M.G., Rumsby B.T. and Newson M.D. 1991. Historic floods and vertical accretion in fine grained alluvium in the Lower Tyne Valley, North East England. In R.D. Hey (ed) Dynamics of gravel bed rivers. Wiley and Sons.

Newbold C., Purseglove J. and Holmes N., 1983. Nature Conservation and River Engineering. NCC.

Petts G.E. 1979. Complex response of river channel morphology subsequent to reservoir construction. Progress in Physical Geography, 3, 329–362.

Petts G.E. 1984. Impounded Rivers. Wiley & Sons.

Richards K. 1992. Rivers : form and process in alluvial channels. Methuen Press.

Roberts C.R. 1989. Flood frequency and urban–induced channel changes: some British examples. In K. Beven and P.A. Carling (eds), Floods: Hydrological, Sedimentological and Geomorphological Implications, 57–82. Wiley & Sons Ltd.

Sear D.A. and Newson M.D., 1992. Sediment and gravel transportation in rivers including the use of gravel traps. NRA project report 232/1/T.

Schumm S.A. 1969. River metamorphosis. Journal of Hydraulic Division American Society of Civil Engineers, 95, 255–273.

Smith S.A. 1989. Sedimentation in a meandering gravel–bed river: the River Tywi, South Wales. Geological Journal, 24, 193–204.

Chapter 6: Suggested Reading

Brookes A.B. 1988. Channelized rivers: perspectives for environmental management. Wiley and Sons.

Lewin J. (ed) 1981. British Rivers. George Allen and Unwin.

Richards K. 1982. Rivers: form and process in alluvial channels. Methuen Press.

7 Rivers: Lowland floods

The Nature of the Problems

Seasonal flooding represents a recurrent problem in many lowland river valleys throughout Britain. This is well illustrated with reference to the River Tay in Scotland, where recent major flood events in 1993 and 1990 have heightened awareness of the risks to local communities. Historical records show that flooding in the Tay catchment has occurred frequently in the past. Indeed records go back as far as the year 1210 when half the town of Perth was said to have been swept away and 'the Kings son and at least 14 others perished'. Over the last 200 years there have been newspaper accounts and records of at least 20 flood events of major significance (Table 7.1). The highest, which occurred in February 1814, was 0.5m higher than the January 1993 flood level at Smeatons Bridge in Perth, and caused extensive inundation which today would have affected a major part of the commercial centre of the city. In contrast to the 1993 floods, however, it was related to river ice which blocked the passage of floodwaters, through the arches of Perth bridge.

The most recent events have generated considerable concern over the flooding issue in Perth, because of the extensive property damage and distress to the local community. Both the 1990 and 1993 floods have been extensively researched by the Tay River Purification Board (e.g. Tay RPB, 1993) and Tayside Regional Council who commissioned consultants to provide a factual record of the floods (Babtie, Shaw and Morton, 1990, 1993; Falconer and Anderson, 1992). These accounts provide much detailed information about the nature and impact of the events; this is summarised below to give a general indication of the type of problems that frequently arise in flood events:

(i)   **February 1990**; exceptionally high river flows between the 4 and 7 February 1990 were caused by prolonged heavy rainfall and very rapid snow melt. This resulted in widespread flooding of land and property in the Tay and Earn Valleys. The flooding also rendered roads and railway lines impassable and left villages and many farms isolated. In total 34m² of land was inundated in the Tay Valley and 8km² in the Earn Valley. Over 42km of agricultural floodbanks were overtopped and embankments were breached at 46 locations (Gilvear and Harrison, 1991). Direct damage costs to both rural and urban areas were estimated to be in excess of £3.2m (Table 7.2), the damage in agricultural areas accounting for 34% of this total, with a major part of this being attributable to reinstatement of floodbanks.

(ii)   **January 1993**; the flood event in Perth and other areas of the Tay catchment between 15 and 18 January 1993 was significantly larger than the 1990 event. For example, at Ballathie gauging station upstream of Perth the peak discharge was 2269m³/s, some 30% higher than the peak flow recorded in 1990. In general river levels were about 0.45 to 0.6m higher than in 1990.

Preliminary estimates indicate that over 1500 properties were affected by flooding, the worst damage being sustained in northern and central areas of Perth, parts of Bridge of Earn and other local communities. Approximately 66km² of land was flooded in the Tay, Isla and Earn valleys.

Perth was severely affected by flooding when a flood embankment to the north of the city was overtopped and breached on 17 January, inundating a housing estate at North Muirton where approximately 780

Table 7.1 Historic flood events in the Perth area (after Babtie, Shaw & Morton, 1990).

Date	Level at Smeaton's Bridge (m)	Weather Conditions	Details
1210	–	Heavy rain, spring tide.	Perth: wooden bridge destroyed. Half of Perth swept away. The King's son and at least 14 others perished.
14 Oct 1621	–	–	Bridge of Perth destroyed. Perth surrounded by water for 5–6 days. Stock losses.
Feb 1733	–	R. Tay at Perth frozen over Jan 1 to Feb 11 on thawing river was chocked by ice.	North Inch, Perth submerged. Five ships thrown ashore.
12 Feb 1814	7.0	Ice blocked passage of flood waters through arches of Perth Bridge at the end of a very severe winter.	High St. (Perth) flooded to the Kings Arms. One ship sunk in harbour, no lives lost.
7 Oct 1847	6.11	Excessive rainfall coupled with a SE wind. Rained from 8pm Tuesday to 5pm Thursday.	Many areas affected in Perth. Inundation and damage to Dunkeld district, Dalguise, Dalmarnock.
19 Jan 1851	5.65	–	–
20 Jan 1853	5.79	–	–
1 Feb 1868	5.90	–	–
7 Feb 1872	–	Two days of constant rain plus strong E winds. Rain dislodged large quantities of snow.	–
7 Feb 1894	5.64	Continuous rain	Perth: North and South Inches, Marshall Pl, Rose Terr, North Port, Lower Commercial St in Bridgend submerged. Much of Tay Valley under water – damage to cultivated land.
31 Jan 1903	5.64	Heavy rains and strong gales. Heavy rain and snow melt from the Grampians. Action of wind on Loch Tay, Loch Rannoch, Loch Tummel raised water levels.	Widespread flooding in Perth
18 Jan 1909	5.52	–	–
19 Aug 1910	5.61	–	–
21 Dec 1912	5.68	Heavy rain, snowmelt. River Tay was abnormally high for two weeks before floods. Incoming tides increased flooding.	Widespread flooding in Perth.
9 May 1913	5.66	–	–
22 Jan 1928	5.77	Wettest January on record. Snowmelt.	Parts of Perth and surrounding areas flooded.
15 Jan 1931	5.49	–	–
15 Jan 1947	5.55	–	–
17 Feb 1950	6.03	Melting snow, heavy rains, high winds.	Parts of Perth and surrounding areas flooded.
5 Nov 1951	5.97	–	–
12 Feb 1962	5.73	–	–

Table 7.1 (cont ...)

Date	Level at Smeaton's Bridge (m)	Weather Conditions	Details
12 Feb 1962	5.73	–	–
31 Jan 1974	5.29	Heavy rain throughout January, snowmelt.	North Muirton flooded. Water supply pump house nearly flooded.
7 Feb 1989	5.07	–	–
5 Feb 1990	5.85	Heavy rain during January snowmelt.	Many areas both in and outside Perth affected.
15 Jan 1993	6.48	Combination of high rainfall and rapid snowmelt.	Over 1500 properties affected in and around Perth.

houses were affected. Due to the sudden failure of the banks many householders did not have sufficient time to remove their possessions to a higher level or to vacate their property. Consequently damage costs were high and emergency services were required to rescue stranded householders. In some areas of the housing estate floodwaters reached a depth of 2m.

Preliminary assessment of the damage costs suggest a figure in excess of £18M, with over £12M of damage within Perth principally due to the flooding of the North Muirton housing estate. The road and rail network was particularly affected by flood damage; costs incurred by the Regional Road Authority were £1.9M and Scotrail amounted to £1.28M. Scottish Hydro Electric was badly affected incurring costs of £0.57M principally due to flooding of property within Perth.

The recent Tay floods also highlight two contrasting mechanisms for flood events:

- **overtopping floods**, as occurred in north Perth during the 1993 event. Here, channel capacity is exceeded by extreme discharges whose probability can be estimated from hydrologic and hydraulic analysis;

- **breach events** following embankment failure as occurred throughout rural areas of the Lower Tay in 1990 and 1993 (Figure 6.1). Such events are the result of complex interactions of river flows and the geotechnical condition of the embankments, and hence, are difficult to predict. Many of

the breaches in 1993 occurred on the outside of bends (35%) and overlay abandoned channels (31%); breaches generally took place in the same location as in past events (Gilvear et al 1994).

Many of the country's lowland rivers follow a pattern of relatively infrequent flood events causing extensive damage to local communities, similar to those described for the Tay. Table 7.3, for example, identifies 30 significant flood events during which floodwaters have remained in floodplain properties for up to 16 days that have occurred on the Thames at Maidenhead over the last 100 years. The 1947 and 1894 events were probably the most damaging; the former has been used as a benchmark level for insurance and planning purposes, with minor modifications made in response to the smaller 1974 flood (Table 7.4). The possible implications of flood events of different severity are outlined in Table 7.5 which identifies the numbers of properties at risk and an indication of the costs that may arise. A repeat of the 1947 flood level (the 56 year return period event) could place over 12,000 people at risk and may lead to £37M damage.

Seasonal flooding is rarely confined to a single river; frequently all the major rivers in a region will overtop their banks, occasionally the flooding can affect many entire regions. In such cases, the damage and devastation can be enormous. Table 7.6 identifies some of the most memorable and destructive floods over the last 125 years; their impacts are similar, with thousands of houses affected, communities cut off by flooded roads and agricultural land, livestock swept away in the swollen rivers, bridges damaged or destroyed, drownings and illness. The costs associated with

Table 7.2 Direct damage for February 1990 flooding in and around Perth (after Falconer and Anderson, 1992).

	Overall Direct Damage Cost		Percentage of Total
1. Agricultural Damage			
Tay catchment			
Floodbank reinstatement	640,000		20%
Other damage	375,000		12%
Earn catchment	50,000		2%
Sub Total		£1,065,000	
2. Building Damage			
Perth	735,000		23%
Outside Perth	425,000		14%
Sub Total		£1,160,000	
3. Public and Other Authorities			
Scotrail	193,000		6%
Scottish Hydro Electric Plc	250,000		8%
Perth & Kinross District Council	80,000		2%
Tayside Regional Council			
Water Services	5,000		–
Roads	381,000		12%
Tay River Purification Board	1,000		–
SSPCA	4,000		–
Tay District Salmon Fisheries	20,000		1%
Sub Total		£3,159,000	29%
Total		£3,159,000	

these events are likely to be in excess of £100M, although it is impossible to gain a true impression of the losses as they are spread throughout the economy and across the nation.

The level of damage associated with particular flood events can vary significantly according to the character of the flood, the land use affected and the action taken by the occupants of the flooded area (Figure 7.1). **Depth** of floodwater is an important factor; damage estimates can be made from depth/damage curves established for various types of property (Figure 7.2). The **duration** of flooding influences the intensity of property damage and is significant in determining the length of the period of disruption. The **velocity of flow** and **sediment load** can also affect the level of damage. Fine deposits, for example, may cause damage to machinery and vehicles, and affect the cost of the clean-up operations.

Flood warnings can have a significant influence on the level of flood damage, especially through preventing loss of life. Table 7.7, for example, highlights the damage reductions associated with flood warning for residential property. However, research has shown that individual responses may be very variable; some did not react with damage-reduction actions, many were sceptical of the warning, others were too old or infirm (Cole and Penning-Rowsell, 1981).

It is recognised that a wide variety of damage can follow flood events. Indeed, several types of flood loss have been identified (Parker et al 1986):

Table 7.3 The Thames at Maidenhead: Flood events since 1982 (after NRA, 1992).

Peak Date	Peak Level (m)	Days Out of Banks	Days in Houses
17 November 1894	24.54	27	9
8 February 1897	23.57	13	4
20 February 1900	23.83	19	8
19 June 1903	23.62	11	6
31 October 1903	23.47	13	4
13 February 1904	23.83	26	14
16 December 1907	23.50	16	4
30 April 1908	23.65	11	4
21 December 1910	23.77	30	6
26 January 1912	23.52	14	4
6 January 1915	24.16	30	15
22 February 1915	23.52	28	6
24 March 1916	23.77	42	16
21 January 1919	23.73	11	5
23 March 1919	23.60	26	15
5 January 1925	24.00	20	9
6 January 1926	23.65	12	5
5 January 1928	23.88	26	7
12 December 1929	24.03	22	8
1 March 1933	23.90	14	7
2 January 1933	23.62	20	5
20 March 1937	23.50	33	8
30 January 1939	23.60	20	4
10 February 1940	23.73	9	6
3 December 1946	23.54	13	4
18 March 1947	24.44	32	14
11 December 1954	23.50	9	4
24 January 1959	23.62	10	5
24 November 1974	23.60	13	3
10 February 1990	23.55	20	10

Notes:

1. Typical Low Flood Level is 22.45m when the river comes out of bank. A major flood is where the level exceeds 24.00m.

2. Flooding of property commences at a level of approximately 23.3m.

Table 7.4 Planning policies for control of residential development in the Thames floodplain around Windsor and Maidenhead.

Residential Development on Land Flooded in 1947 (excluding land flooded in 1974).

- new developments not to exceed 10% of site area;

- residential density less than 35 habitable rooms per acre;

- extensions to existing property not to exceed 300 sq. ft.;

- all property searches should be notified of flood risk in the area of the 1947 flood.

Residential Development on Land Flooded in 1974.

- no new development in the 1974 flood area;

- no sub-division of existing development sites;

- extensions to existing property not to exceed 30 sq. ft.;

- no conversion from non-residential to residential use;

- no development that raises garden levels;

- ground floors in new dwellings should be at least 15cm above the 1947 flood level.

Summarised from: Royal Borough of Windsor and Maidenhead, 1982.

Table 7.5 Number of residential properties at risk from various floods in the Maidenhead area and estimated financial cost (after NRA, 1992).

County/ Parish	Number of residential properties at risk for particular return period floods						Total properties in Area
	5	9	25	56	101	204	
Berkshire	101	683	2572	4364	5382	6015	6982
Bray	0	5	85	215	269	320	572
Cookham	5	11	34	75	121	164	274
Eton	0	0	167	505	891	1140	1252
Maidenhead	96	307	1311	2123	2517	2707	2826
Slough	0	0	25	62	72	82	102
Windsor	0	360	950	1384	1512	1602	1956
Buckinghamshire	26	106	231	298	369	415	513
Burnham	0	0	0	1	1	1	5
Dorney	1	4	30	60	96	119	168
Taplow	8	82	164	197	224	237	244
Wooburn	17	20	37	40	48	58	96
Total Number	127	789	2803	4662	5751	6430	7495
Total People Affected	338	2099	7456	12401	14500	17104	19936
Total Event Damage £M	0.66	2.22	11.96	36.68	57.69	88.88	

Table 7.6 A selection of the major river flood events in Britain since 1875.

Date	Area Affected	Comment
26–27 December 1979	South Wales	Widespread flooding; 2 dead in Rhydycar, 2 dead in Cardiff, thousands evacuated. Hundreds of homes flooded, bridges damaged, roads and railways affected.
10 July 1968	South West England	Extensive river flooding throughout the region, up to 9 dead and increased incidence of illness and death in flooded areas of Bristol in years following the flood. Thousands of houses flooded, cattle drowned, roads washed away.
10 December 1965	Midlands, South East England, Yorkshire	Widespread flooding, communities cut off, roads damaged, buildings flooded, deaths reported.
4 December 1960	South Wales, South West England, Midlands	Extensive flood damage, thousands of houses flooded, bridges swept away, roads impassable; Central Bath blacked out because of a power failure. In Exeter area 3000 homes affected.
March 1947	South Wales, Midlands, South East England, East Anglia Yorkshire	The most severe floods of the century; enormous variety of damage to numerous communities throughout Britain. Water supplies affected in the Lea Valley; flood waters up to roof level in Dunham; millions of pounds of damage.
16 November 1894	South East England Midlands, South West England	One of the largest flood events, especially on the Thames, Wye, Severn and Bristol Avon; damage and destruction to houses and businesses, sheep and cattle washed away, communities cut off.
8 March 1889	Midlands South West England	Extensive flooding of the Severn, Trent, Warwickshire Avon and many other rivers; thousands of houses damaged, agricultural land flooded, bridges lost.
December 1876–January 1877	Midlands South East England South West England	Extensive flooding of the Thames, Severn, Trent and many others. Numerous towns and villages badly affected, thousands of houses damaged, land flooded, communications disrupted.
14–21 July 1875	Midlands South Wales South East England	At least 10 drowned in widespread flooding over many days; towns and villages badly damaged, land flooded, cattle and sheep swept away.

Table 7.7 Generalised data on the benefits of flood warnings measured as damage reduced (from Cole and Penning-Rowsell, 1981).

Residential Property					
Depth of Flooding (m)	Total Potential Damage £	Estimated Average Damage Reduction with Flood Warnings			
		Up to 2 hour Warning		2–4 hour Warning	
		£	%	£	%
1.2	2220	400	18	550	25
0.9	2030	400	20	550	27
0.6	1740	300	17	450	26
0.3	908	450	50	500	55
01	338	150	44	150	44

Figure 7.1 The relationship between flood character and risk (after Green et al, 1983).

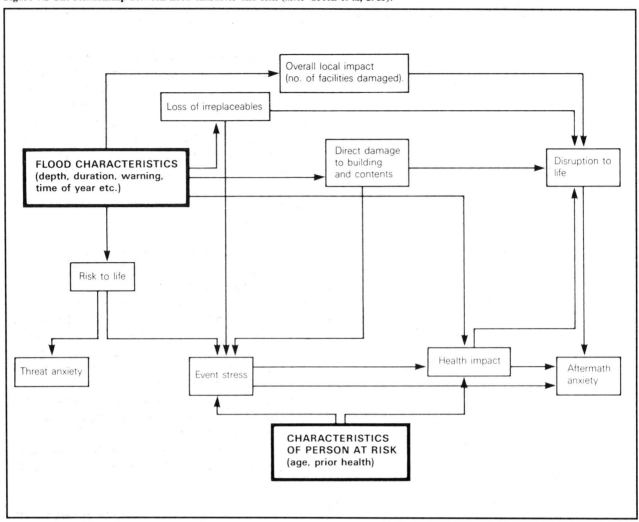

(i)   **direct damages** caused by the physical contact of floodwater with properties and their contents;

(ii)   **indirect damages** arising as a consequence of direct damage, including: traffic disruption, loss in production, evacuation costs, etc;

(iii)   **intangible damages** ranging from anxiety and stress to ill health related to the general inconvenience caused by the event.

The **Flood Hazard Research Centre** at Middlesex University has carried out a number of investigations into the effects of floods, many of which demonstrate that indirect and intangible damages are often underestimated (e.g. Green and Penning–Rowsell, 1989; Parker, 1991). Their detailed research into the impact of the 29 July 1987 floods at Waltham Abbey and Thornwood, Essex provides an excellent catalogue of damage

and can be seen as indicative of the level of costs that can occur (Tunstall and Bossman–Aggrey, 1988). The floods occurred in early evening as a result of an intense afternoon thunderstorm, with two small streams not capable of carrying the unusually large discharge. Water entered over 100 homes which were flooded to an average depth of 0.25m and 0.52m at Waltham Abbey and Thornwood, respectively. Table 7.8 gives details of the amounts spent to make good the structural and property damage, which included:

●   floor boards needed treatment with fungicides;
●   redecoration;
●   carpets and furniture ruined;
●   electricity sockets replaced;
●   gardens ruined;
●   shed contents damaged.

The intangible effects of the floods included general disruption, with some homeowners evacuated for many weeks after the event. The

Figure 7.2 The relationship between flood depth and damage for various land uses (after Penning-Rowsell et al, 1986).

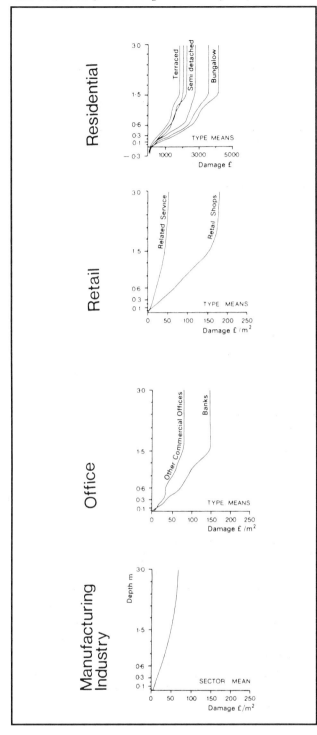

- anxiety and nervousness (16%);
- colds, viral infections, etc. (14%);
- trouble sleeping (9%).

The health effects of floods were examined in detail by Bennett (1970) who compared the health record of those affected by the 1968 Bristol floods with a control group outside the flooded area. Interviews with flood victims showed a marked increase in physical ill-health amongst males and psychiatric problems in females. The numbers of hospital referrals more than doubled in the year after the floods; the mortality rate for those affected rose by 50%, especially for males of age group 45-64.

In many areas the scale of potential problems associated with floods have increased. This, in part, reflects a changing attitude to flood risks. In the past, floodplain residents were generally aware of the potential problems and took precautions as and when necessary, moving furniture and valuables upstairs when river levels rose. This voluntary acceptance of the risk has, to an extent, been replaced by a lack of appreciation of flood problems. High residential mobility in new housing estates has contributed to this situation, as new occupants do not have the accumulated flood experience of residents in more settled floodplain communities (Smith and Tobin, 1979; Parker and Penning-Rowsell, 1982).

The potential problems have also been heightened by the rise in property values and ownership of expensive household goods such as televisions, washing machines, fitted kitchens, carpets etc. all of which are very vulnerable to flood damage. Suleman et al. (1988) suggest that the **damage potential** to floods rose by over 50% for short duration and 100% for long duration events, between 1977 and 1987. This trend has been reinforced by the continued expansion of housing development in floodplain areas. In Maidenhead, for example, flood damage potential has risen dramatically since the major 1947 event described earlier. At that time there were 1400 properties at risk, by 1990 the figure had risen to 3,558; estimated flood damage costs have risen from £1.3M in 1947 to a potential figure of £19M by 1990 (Parker, 1991).

The significance of floodplain encroachment is well illustrated by the example of Datchet, a village on the Thames, to the west of Windsor. Despite the potential for large-scale flood damage, over 600 new residential properties were permitted within the area affected by the 1947 Thames flood,

clean-up operation also created problems as delays were frequently experienced in getting building and decoration work done; on average the affected people took between 62-124 days to get their homes back to normal. The stress, worry and health effects were many and varied, including:

- mental stress, marital strain (34% of residents affected);
- depression (18%);

	Waltham Abbey (65 Houses Flooded)	Thornwood (38 Houses Flooded)
Repairs to structure and decorations	£152,640	£110,580
Repairs and replacement of furniture and contents	£94,130	£123,840
Damage to garages, gardens, sheds etc.	£21,820	£5,690
Damage to cars	£7,220	£4,310
Extra heating costs	£1,650	£2,150
Total	£277,460	£246,570

between 1974 and 1984 (Figure 7.3; Neal and Parker, 1988). One of the reasons for the development pressure in and around the village has been the presence of the surrounding **greenbelt** which has had the effect of diverting development onto the floodplain.

Flood events are frequently accompanied by erosion and deposition. Indeed, bank erosion may contribute to the flood problems through causing embankment failure, as was the case in the Lower Tay in both 1990 and 1993. Deposition of fine sediments across the floodplain can be an important factor in determining the severity of damage associated with an event. Prolonged floods are often accompanied by the deposition of large quantities of sediment within a channel. For example, flooding on the River Trent can lead to serious navigation restrictions as up to 1.8m of sediment may build up in a single week of prolonged flooding.

The Causes of Lowland Floods

Although the impacts of flooding are all too familiar, floods are difficult to define precisely as they are viewed differently by different people. To a hydrologist a flood is merely an extreme discharge from a data series of river flows that range from droughts to very high streamflows. From a land management perspective, the flows that exceed the channel capacity and lead to overtopping of the banks are the most important. However, channel capacity is not constant along a river or through time,

making it difficult to equate flooding with a particular discharge threshold. In addition, the resulting damage is central to defining the character of a flood. Many areas vulnerable to frequent flooding are in rural areas where the damage is less severe than in built up areas. Thus, although the largest floods recorded in Great Britain tend to occur in Scotland and Northern England (Table 7.9), the most memorable events have occurred in the densely populated southern England and Wales.

The cause of a flood event is self evident; too much water for the channel to carry. The factors that influence flood behaviour can, therefore, range from regional controls on the supply of water to local influences on channel capacity, and include:

- climate;
- catchment characteristics;
- land management;
- site factors.

Excessively heavy and prolonged rainfall is the most common cause of flooding. These conditions are frequently associated with winter cyclones or summer thunderstorms. As described in Chapter 2, the west is wetter than the east of Britain (Figure 2.13), with the number of raindays increasing from 165 in the south east to over 230 in the Scottish Highlands. The variation in mean annual discharge (Figure 7.4) is remarkably similar to the average annual rainfall (Figure 2.13), emphasising the importance of climate to determining flood behaviour. The pattern of extreme rainfall events is much more complex (Figure 2.15), indicating that flood triggering events can occur throughout Britain, albeit with much longer return periods away from upland areas.

The occurrence of a large rainfall event does not in itself ensure that a flood will follow. Much of the rain may be intercepted or lost by evaporation; the rate of runoff will depend on the infiltration capacity of the surface materials. Antecedent rainfall conditions play an important role in **preparing** a catchment for a flood event, and frequently dictating the severity of flooding. The rainfall during the period preceding a flood can saturate hillslopes, and may lead to greater and faster runoff in subsequent storms. This effect can lead to the marked seasonal variation in flood response to similar rainfall events in the same catchment. Harvey (1971) described seasonal flood behaviour in the Ter catchment in eastern England, noting that summer floods required higher rainfall intensities than winter floods when even low rainfall events could produce substantial floods. Frozen ground is another important preparatory factor, reducing infiltration and

Figure 7.3 The encroachment of floodplain development at Datchet, Berkshire (after Neal and Parker, 1988).

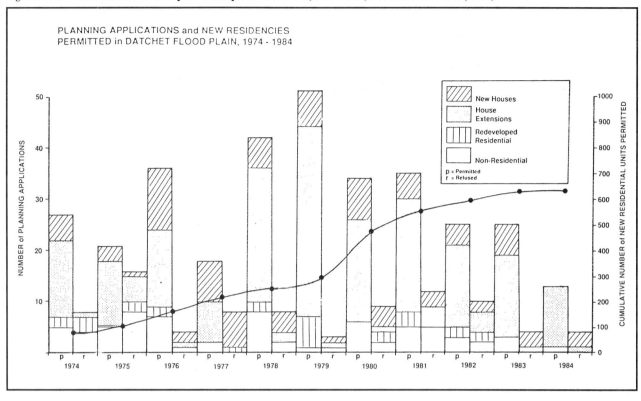

Figure 7.4 Estimated mean annual flood across Great Britain (after NERC, 1975).

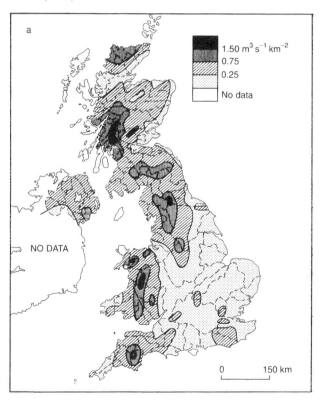

dramatically increasing runoff.

Melting snow can be an important cause of seasonal flooding, especially in the Scottish Highlands and other upland areas. The most severe floods tend to follow periods of heavy snowfall when temperatures rise rapidly after a long cold period, as happened in March 1947. In that year much of the country had been covered by snow from 27 January to 13 March, with 5m deep drifts in places. On 10 March warm air arrived in south west England with rain (snow melts at around 65mm/day in warm air; 250mm/day in warm air and rain). By the 11th, vast areas from Somerset to Kent were inundated, with severe flooding on the Thames, Avon, Medway and Lea. Over the next few days the warm air spread to the north and east. The Severn and Wye flooded, leaving Herefordshire almost isolated. By the 13th several Norfolk rivers were in spate. Severe gales and rain on the 16th led to flooding of the Great and Little Ouses, and the Welland at Spalding. The Trent overtopped its banks on the 18th, flooding hundreds of floodplain homes up to first floor level. On the Humber, Gainsborough was under water. In Yorkshire, the Wharfe, Derwent and Aire were all swollen with floodwater from the melting snow.

Catchment characteristics have an important bearing on flood behaviour, although the number of potential variables relating to drainage network, channel characteristics and underlying materials is enormous.

Table 7.9 Peak discharge estimates for some major floods since 1795 (after Acreman, 1989).

Date	Site	Peak Flow (m³/s)	Area (km²)
January 1993	Tay, Perth	2269	4587
August 1970	Divie & Dorback, Morayshite	1939	365
August 1829	Dee, (Woodend) Aberdeenshire	1900	1370
January 1849	Ness, Inverness	1700	1792
January 1974	Tay, Ballathie	1570	4587
October 1967	Esk, Netherby	1545	842
January 1982	Tweed, Norham	1518	4390
November 1951	Tay, Caputh	1481	3211
February 1795	Trent, Nottingham	1416	7490
January 1982	Tweed, Sprouston	1409	3330
October 1987	Twyl, Carmarthen	1378	1088
January 1962	Nith, Friars Carse	1274	799

The **UK Flood Study** used a number of regional factors, together with climate, to estimate the size of the mean annual flood (NERC, 1975); these factors were drainage area, stream frequency, stream channel slope, a soil index, a measure of rainfall excess and the proportion of the catchment occupied by lakes or reservoirs. The Flood Study also identified that the relationship between mean annual flood and the flood of a particular return period varied across the country. Ten distinct regions were recognised for regional flood predictive purposes (Figure 7.5) which indicate that in lowland areas, extreme events such as a 100 year flood may be 3 to 4 times the mean annual flood. In upland areas a similar return period event may be only twice the size of the annual flood. This pattern is largely a function of catchment character and has major implications for the size of defence works needed to protect against floods of a given return period.

Land management and urbanisation can have a significant effect on flood behaviour, especially urbanisation (described below under the effects of development). Indeed, the Flood Study identified that urbanisation was one of the main factors influencing the time to peak discharge i.e. speed of response to a rainfall event. Although much has been written about the effects of drainage for agriculture and forestry in upland areas (see Chapter 5), this factor was not considered significant influencing the seasonal flood behaviour of lowland areas (NERC, 1975).

At the site level, the capacity of the channel to carry large discharges can be affected by a wide range of factors. As was described in Chapter 5, debris trapped behind bridges or culverts can cause temporary dams, especially on small rivers. Rubbish and debris can be allowed to build up within channels, reducing the ability to cope with flood flows (a factor in the Waltham Abbey and Thornwood floods of 1987). Deposition of large amounts of sediment within the channel during a flood event can significantly reduce bankfill capacity and can lead to further flooding unless the channel is cleared, for example, by dredging. This can be an important factor in many Scottish rivers such as the Spey and the River Nith at Dumfries. Landslides can block the channel, causing a build up of water behind the dam. Although this is not a common occurrence in Britain, one recent example took place on the River Colne in 1992 when a 21m high sludge tip failed at a sewage works near Huddersfield. The slide blocked the river over a 150–200m length, raising fears of flooding which were abated by digging a channel to divert the river around the slide (Pellymounter, 1992).

Flood problems can also occur when the flow in tributary streams, drainage channels or sewers cannot discharge into a swollen river, causing the water to back up. This was a factor in the flooding of many city centre properties in Perth during the 1990 and 1993 floods, when the sewers and small burns could not drain into the Tay.

Figure 7.5 The relationship between return period and the size of flood event, as related to the mean annual flood, for 10 regions in Great Britain (after NERC, 1975).

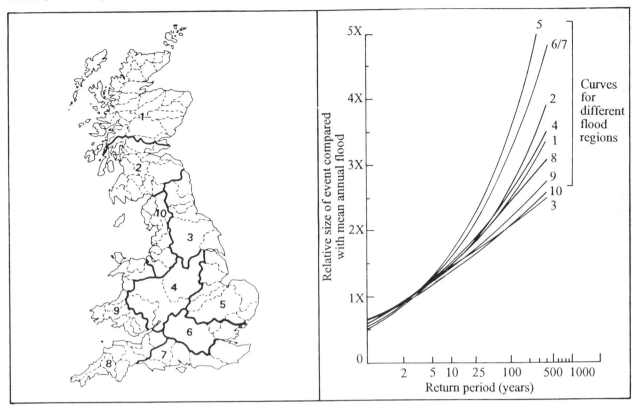

Many large flows are regulated by reservoirs or contained by riverside defences such as embankments; breaches or failures of these man-made structures can cause particularly severe floods. Dam disasters are rare in Great Britain; the last failure resulting in loss of life was in 1925 (see Chapter 5). However, breaching of flood defences is much more frequent, especially in agricultural areas where the defences may be little more than earth banks raised by the farmer to protect against low return period flood events. Research on the recent Tay floods has shown that although past breaches may have been repaired, these sites remain highly vulnerable locations (Gilvear et al, 1994). Floods may also occur as a result of breaches of canal embankments, as occurred in Coventry in 1978, when millions of gallons of water flooded about 60 homes in the city, forcing people to evacuate, (Perry, 1981).

The Identification of Vulnerable Areas

Floods occur on floodplains, the low-lying land adjacent to streams and rivers. These features are readily identifiable as their lateral extent is normally marked by a distinct break of slope where the floodplain abuts a hillslope. At a national level, the extent of land potentially vulnerable to seasonal flooding can be derived from the 1:250,000 scale National Soil Maps (see Chapter 3) which show the extent of **floodplain soils**. These areas are highlighted on the accompanying 1:625,000 scale thematic maps.

The floodplain, however, is often wider than the extra channel capacity needed to cope with flood flows. Many streams are "misfits" with their board floodplains more a reflection of past wetter periods of climate than present flood behaviour. At a local level, therefore, more precise information is needed and the identification of vulnerable areas usually involves defining the extent of major flood events, such as the 1947 floods on the Thames. In England and Wales the compilation of this information was the responsibility of the water authorities and, since privatisation, the NRA who assumed responsibility for carrying out surveys of flood risk areas and passing the results on to local planning authorities. There is no equivalent to the NRA in Scotland, but relevant information about the extent of past flooding in some urban areas is likely to have been compiled by the Regional Councils.

The areas at risk can, of course, be modified through the construction of defence works. As a result, a discharge comparable to a historical flood event may not necessarily flood the same area. It is important to stress, however, that the construction of flood defence only **reduces** the risk. It cannot **eliminate** it. Much depends on the design life of the structure and the degree of risk that is acceptable. It should be noted that a scheme designed to cope with the 100 year event (Figure 7.6) would only have a 45% probability of not failing within 100 years. Thus, the tendency for increased investment and density of development behind defences may only lead to higher losses when, inevitably, floods occur that are larger than that for which it was designed to provide protection.

The recent Tay floods have shown, however, that flood embankments on the outside of bends or crossing abandoned channels are particularly vulnerable to breaching. In this context, Gilvear et (1994) have suggested a number of floodplain management solutions:

- relocate embankments away from critical locations;

- provide adequate spillways to allow controlled inundation;

Figure 7.6 The relationship between the standard of protection (as expressed by a return period event), the design lifetime of structure & the chance of failure during this period

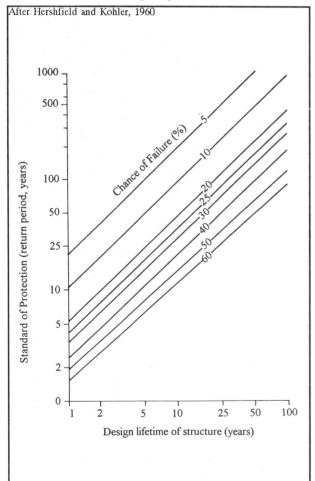

- consider land use management changes through, for example, set–aside (Gilvear 1993).

The Effects of Development

Some of the effects of urban development on flood behaviour have already been described in the context of a temporary increase in soil erosion (Chapter 3), together with the increase in quantity and rate of runoff and its impact on flash flooding (Chapter 5) and channel form (Chapter 6). The nature of seasonal flooding of lowland areas can be affected by two key factors: urban drainage systems and the reduction in flood storage capacity when development encroaches onto floodplains.

The implications of urban drainage are complicated by the variability of existing systems (Figure 7.7), involving the combined stormwater and foulwater systems favoured by early municipal engineers and the modern separate systems. The nature of flood related problems depends on how the drainage system is connected to the river system (Roberts, 1989). Older developments with combined systems tend to generate less hydrological change because the majority of both surface and foul water enters the river only after passing through a sewage treatment plant. However, these systems usually incorporate storm water outflows to protect the treatment works from the impact of storm events. These outflows discharge to watercourses and may lead to flooding with sewage contaminated water. By contrast, modern suburbs tend to have large diameter drains, allowing rapid runoff direct to the nearest watercourse. This can lead to an increase in flows in the channel and a quicker response to rainfall events. The size of change is related to the extent of development and the return period of a particular event; small events are affected most. A one year return period event can be enhanced ten times for 40% urbanisation, whereas the two year event is only doubled or trebled (Hollis, 1975). At both Stevenage and Skelmersdale, for example, Knight (1979) found that the mean annual flood was about 2.5 times its former size after urbanisation. The modification of more extreme events is less, probably because during these events the catchment may already be responding as if impermeable because extensive areas are saturated.

Floodplain storage is an important mechanism by which a flood wave can be dampened as it travels down a reach, reducing the peak of the flood but extending its duration (i.e. the flood is spread out

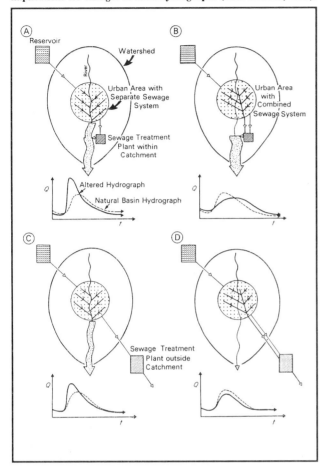

over a longer period and, hence, is less severe). Figure 7.8 demonstrates for the Lower Tees, between Broken Scar and Low Moor, how, beyond a threshold discharge, there is a clear reduction in the speed that a floodwave travels; the wave travel time along the reach increases from 5.6 hours at around $200m^3/s$ to 8.7 hours at the highest observed flows (Archer, 1989). Storage can also have a significant effect on discharge; the twenty year flood at Low Moor is frequently around $200m^3/s$ less than the same flood event at Broken Scar because of the extensive floodplain storage along the reach (Figure 7.9). Flooding of rural areas can be important for the relief of flooding in urban areas downstream, although it can be detrimental to agricultural interests. In these circumstances, the construction of low earth embankments to provide against low-return period floods can reduce the frequency of inundation of agricultural land whilst maintaining flood storage for higher flows which are allowed to overtop the banks.

Expansion of development onto floodplains can significantly reduce the storage capacity of a river system. Permeable surfaces are replaced by impermeable ones, reducing the potential loss of floodwater through infiltration into the ground. The construction of flood defence embankments to protect vulnerable communities are designed to prevent floodplains acting as stores and can, therefore, create problems in downstream reaches where the loss of flood wave attenuation can lead to more severe flooding than may have previously been the case.

The Significance for Conservation

Flooding of low lying land adjacent to rivers can be of obvious benefit to many river corridor habitats such as bogs, marshes, ox-bow lakes and backwaters. Conservation sites dependent on a degree of regular flooding are widespread throughout Britain. The value of these river corridor habitats has been increasingly recognised, especially since the Wildlife and Countryside Act 1981 required drainage authorities in England and Wales to further conservation. Indeed, it has been this conflict between flood defence issues and conservation that has been central to river management; considerable environmental "damage" probably has resulted from the construction of some defence works.

However, there have been major changes in river management approaches. The Water Act 1989 and with provisions re-enacted in the Water Resources Act 1991 and Land Drainage Act 1991 placed even greater responsibility on drainage authorities. Environmental Assessment is required for all improvement works which might have an adverse environmental impact; the NRA has an additional power to 'promote' conservation provided it is consistent with its other duties. Guidance documents on sensitive management have now been prepared by Central Government, including:

Figure 7.8 Flood wave speed and travel time through the Tees at Broken Scar (after Archer, 1989).

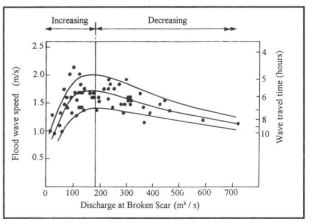

Figure 7.9 Floodplain storage on the Tees (after Archer, 1989).

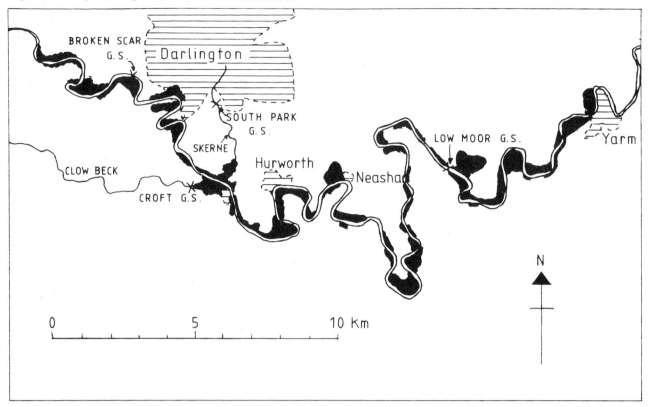

DoE/MAFF/WO, 1989 – Code of Practice on Conservation, Access and Recreation.

DoE/WO, 1989 – Environmental Assessment. A guide to procedures.

MAFF/DoE/WO, 1991 – Conservation Guidelines for Drainage Authorities.

MAFF/English Nature/NRA 1992 – Environmental procedures for inland flood defence works: a guide for Managers and Decision-makers in the NRA, IDBs and Local Authorities.

At the same time the need to integrate nature conservation interests with management of IDB drainage channel maintenance was recognised (Newbold, et al, 1989). An indication of the needs of adequate standards of service and having sympathy for the needs of the environment is illustrated by the IWEM Water Practice Manual (Brandon; 1988) which devotes over one seventh of its contents to sensitive river management techniques which benefit the environment. MAFF guidance on inland flood defence works (MAFF/EN/NRA, 1992) embodies best practices which should be followed before undertaking inland works.

An updated version of the "Rivers and Wildlife Handbook" is currently in preparation, drawing on experience from a range of bodies including NRA,

RSPB and RSNC (Holmes, 1993). Information on how and why rivers affect floodplains (and vice-versa) is given in addition to guidance on the law relating to rare species. Technical description of management activities and their effects on river and floodplain habitats, flora and fauna are also included. Areas covered include:

- Bank and Floodbank Vegetation Management and Establishment;
- Aquatic Vegetation Control and Management;
- Riparian Tree and Shrub Management;
- In-stream Habitat Enhancements;
- River Management – including: dredging and re-sectioning, multi-stage channels, re-alignment and by-passes, deflectors and revetments, floodbanks, weirs and sluices, and flood storage.

Summary: The Significance for Planning and Development

Lowland rivers regularly overtop their banks and inundate the adjacent floodplains, especially following periods of heavy rainfall or during snow melt. Enormous levels of damage and disruption can occur where development has occurred on floodplains. For example, the 1993 Perth floods resulted in around £18M of damage; a repeat of the

1947 flood level on the Thames could lead to nearly £40M of damage in the Maidenhead area alone.

The severity of damage is generally related to the depth and duration of flooding, together with the degree of forewarning provided to the residents of affected areas. Floods are traumatic events, creating major disruption to lives for months after the event. Indeed, studies have shown that physical and psychiatric health problems frequently follow major flood events. Flooding events are often accompanied by large scale in-channel deposition, leading to restrictions to navigation and requiring emergency dredging works.

Although the causes of flooding are self evident – too much water for the channel to carry –they should not be regarded as purely a natural occurrence. Land use and development can have significant effects on the flood character of an area in a variety of ways, most notably through increases in runoff generation and reduction in floodplain storage. Encroachment of development onto floodplains increases the potential levels of damage when floods occur. In addition, rises in property values and ownership of expensive household goods have tended to increase the flood damage potential. In the context, it has been suggested that the damage potential has risen by over 100% since the mid 1970s. The risks can, of course, be modified through the construction of defence works. However, they can only reduce the risk and not eliminate it. The tendency for increased investment and density of development behind defences may only lead to higher losses when, inevitably larger floods occur.

Floods are not, however, wholly detrimental. Many valued habitats such as bogs, marshes, ox-bow lakes and backwaters are dependent on a degree of regular flooding. It should be appreciated that a reduction in flood risk in one area may have significant environmental effects elsewhere, as emphasised by the recent MAFF Conservation Guidelines for Drainage Authorities (MAFF/DOE/WO, 1991).

Flood risk can be an important constraint to development. Indeed, development on floodplains can lead to danger to life, damage to property and wasteful expenditure on remedial works at the development site and elsewhere. Local planning authorities can minimise the effects of flooding through:

- guiding development away from flood risk areas;

- restricting development that could increase the risk of flooding, either through the generation of additional runoff or the reduction in floodplain storage;

- restricting development that would interfere with the ability of drainage operators to carry out flood control works and maintenance;

- ensuring that land allocated for development in flood risk areas can be satisfactorily protected by the developer without affecting nature conservation or other environmental interests.

In England and Wales, guidance to local planning authorities and others is provided in Circular 30/92 **Development and Flood Risk** (DoE, 1992). No comparable advice has been issued by the Scottish Office (NB: draft guidance was issued by the Scottish Office in March 1995).

Chapter 7: References

Acreman M.C. 1989. Extreme historical UK floods and maximum flood estimation. Journal of Institution of Water and Environmental Management, 3, 404–412.
Archer D.R. 1989. Flood wave attenuation due to channel and floodplain storage and effects on flood frequency. In P. Carling and K. Bevan (eds) Floods: Hydrological, Sedimentological and Geomorphological Implications, 37–46. Wiley.
Babtie, Shaw and Morton, 1990. Flooding in the River Tay Catchment. Report to Tayside Regional Council.
Babtie, Shaw and Morton, 1993. Flooding in the Tay Catchment – January 1993. Report to Tayside Regional Council.
Bennet G. 1970. Bristol floods 1968: controlled survey of effects on health of local community disaster. British Medical Journal, 3, 454–458.
Brandon T.W. (ed) 1988. Water practice manuals. River Engineering Part II. Structures and Coastal Defence Works. IWEM London.
Cole G and Penning-Rowsell E.C. 1981. The place of economic evaluation in determining the scale of flood alleviation works. In Flood Studies Report – 5 years on. ICE, 143–151.
Department of the Environment 1992. Development and Flood Risk. Circular 30/92 (MAFF Circular FD1/92; Welsh Office Circular 68/92). HMSO.
Falconer R.H. and Anderson J.L. 1992. The February 1990 flood on the River Tay and

subsequent implementation of a flood warning system. IWEM Joint Meeting, Perth 1992.

Gilvear D.J., 1993. River management and conservation issues on formerly braided river systems: the case of the River Tay, Scotland. In C. Bristow and J. Best (eds.) Braided Rivers. Geological Society Special Publication, 75, 231–240.

Gilvear D.J. and Harrison D.J., 1991. Change and the significance of floodplain stratigraphy: 1990 flood event, lower Tay, Scotland. Earth Surface Processes and Landforms, 16, 753–762.

Gilvear D.J., Davies J.R. and Winterbottom S.J. 1994. Mechanisms of Floodbank failure during large flood events on the rivers Tay and Earn, Scotland. Quarterly Journal of Engineering Geology 27, 319 – 332.

Green C.H. and Penning–Rowsell E.C. 1989. Flooding and the quantification of "intangibles". Journal Institution of Water and Environmental Management, 3, 27–30.

Green C.H., Parker D.J. and Emery P.J., 1983. The real cost of flooding to households: intangible costs. Geography and Planning Paper No. 12, Middlesex University.

Harvey A.M. 1971. Seasonal flood behaviour in a clay catchment. Journal Hydrology, 12, 129–144.

Hershfield D M and Kohler M A 1960. An empirical appraisal of the Gumbel extreme–value procedure. Journal of Geophysical Research 65, 1737–1746.

Hollis G.E. 1975. The effect of urbanisation on floods of different recurrence intervals. Water Resources Research 11, 431–435.

Holmes N. 1993. Opportunities and practice: inland works. Proc. 1993. MAFF Conference of River and Coastal Engineers, Loughborough.

Knight C.R. 1979. Urbanisation and natural stream channel morphology: the case of two English new towns. In G.E. Hollis (ed) Man's impact of the hydrological cycle in the United Kingdom, 181–198. Geobooks.

MAFF/DoE/Welsh Office 1991. Conservation Guidelines for drainage authorities.

Neal J. and Parker D.J. 1988. Floodplain encroachment: a case study of Datchet, UK. School of Geography and Planning Paper No. 22. Middlesex University.

NERC 1975. Flood Studies Report. Institute of Hydrology, Wallingford.

Newbold C., Purseglove J. and Holmes N. 1983. Nature conservation and river engineering. NCC Peterborough.

NRA 1992. Proof of Evidence: Maidenhead, Windsor and Eton Flood Alleviaion Scheme. NRA Thames Region.

Parker D.J. 1991. Flood disasters in Britain: lessons from flood hazard research. Flood Hazard Research Centre Paper. Middlesex University.

Parker D.J., Green C.H. and Thompson P.M. 1986. Urban flood protection benefits: a project appraisal guide. Farnborough: The Technical Press.

Parker D.J. and Penning–Rowsell E.C. 1982. Flood risk in the urban environment. In D.T. Herbert and R.J. Johnson (eds) Geography and the Urban Environment, 201–239. Wiley.

Pellymounter D. 1992. The NRA response to the landslip on the River Colne, Huddersfield. Proc. 1992 MAFF Conference of River and Coastal Engineers, Loughborough.

Penning–Rowsell E.C. and Chatterton J.B. 1977. The benefits of flood alleviation: a manual of assessment techniques. Saxon House.

Perry A.H. 1981. Environmental hazards in the British Isles. George Allen and Unwin.

Roberts C.R. 1989. Flood frequency and urban–induced channel change : some British examples. In P. Carling and K. Bevan (eds) Floods : Hydrological, Sedimentological and Geomorphological Implications, 57–82. Wiley and Sons.

Royal Borough of Windsor and Maidenhead, 1982. Policy for the control of residential development in the flood plains of the River Thames and the River Colne.

Smith K. and Tobin B.A. 1979. Human adjustment to the flood hazard. Longman.

Suleman M.S., N'Jai A., Green C.H. and Penning–Rowsell E.C., 1988. Potential flood damage data : a major update. Flood Hazard Research Centre, Middlesex University.

Tunstall S. and Bossman–Aggrey P. 1988. Waltham Abbey and Thornwood, Essex. Flood Hazard Research Centre, Middlesex University.

Chapter 7: Suggested Reading

Newson M.D. 1975. Flooding and flood hazard in the United Kingdom. Oxford University Press.

Penning–Rowsell E.C., Parker D.J. and Harding D.M. 1986. Floods and drainage. George Allen and Unwin.

Smith K. and Tobin G.A. 1979. Human adjustment to the flood hazard. Longman.

Ward R.C. 1978. Floods: a geographical perspective. Macmillan.

8 Estuaries: Sedimentation

The Nature of the Problems

Estuaries provide natural harbours and transport routes and, consequently, have been the focus of considerable port activity. As an island nation, ports and harbours have always been central for trade with overseas markets, transport, national security and the fishing industry. There are over 300 harbours around the coast. Sedimentation within estuaries, however, causes major problems for many British ports with periodic dredging required to keep the port facilities open to vessels. Two main types of dredging operations occur:

(i)   **maintenance dredging**; routine dredging to keep the navigable channel open;

(ii)   **capital dredging**; dredging of new channels or berths, deepening or changing existing channels to ensure that the port can be competitive by handling larger vessels.

Most port and harbour facilities require dredging. The exceptions are mainly the estuaries where the use is mainly recreational, such as the Artro in West Wales. Table 8.1 highlights the range of problems that have been reported by a sample of port authorities. Dredging requirements at both Harwich and on the Tees cost over £1M a year; considerable volumes of material are removed from the estuary bed – 1.3M tonnes a year from the Mersey alone; King's Lynn and Wisbech are affected by shifting channels which have to be regularly resurveyed.

The costs of managing the sedimentation problems can be a significant factor in determining the efficiency of a port. Indeed, rising costs and increased competition has led to a trend for the relocation of many port facilities downstream, closer to the river mouth, where the channels are deeper and dredging requirements are less of a constraint. Felixstowe and Tilbury are examples of this pattern.

Port authorities may also be faced with considerable difficulties in disposing of the dredged spoil. In the past many ports have used the material for land reclamation schemes. At Southampton, for example, frequent dredging of the upper estuary produced $500,000m^3$ of spoil a year which was used to fill saltmarshes on the south west side of the estuary (Coughlan, 1979). However, the practicality and cost of land–based disposal options means that much of the dredged material is currently dumped at sea (Table 8.2). This is controlled under licences issued by the relevant fisheries department, under the Food and Environment Protection Act 1985, with the objectives of ensuring that no harm will be caused to the marine environment, nor will it interfere with other interests. Such licences will only be granted to port authorities where the licensing authority is satisfied that these objectives will be met and that there is no practical alternative disposal option available.

The consideration of alternatives to disposal at dump sites at sea can be problematic. Ash (1994) has identified four broad categories for the **beneficial use** of dredgings, including coastal defence (e.g. beach nourishment), habitat creation, land reclamation for development and retention within the coastal system to maintain the sediment budget. Table 8.3 highlights the beneficial uses of dredgings from the expansion of the Port of Felixstowe in the early 1990's, with $760,000m^3$ of coarse material used in NRA coastal defence schemes and $400,000m^3$ used for habitat creation. However, the majority of dredged material is fine, the properties of which are poorly understood and for which beneficial uses are harder to find, especially when the sediment is contaminated with heavy metal pollutants.

Table 8.1 Examples of deposition problems and annual dredging costs from selected British ports.

Port	Annual Cost	Comment
Aberdeen	£200,000	Ingress of sand at the entrance and deposition of silt in the berths.
Berwick	n/s	Deposition of silt and gravel at the entrance to Tweed dock; dredging occasional.
Bridlington	n/s	35,000 tonnes of silt and mud removed a year.
Clydeport	n/s	Dredging is a major problem, amounts include: 1989 – 282,700 tonnes 1990 – 188,400 tonnes 1991 – 251,500 tonnes 1992 – 130,000 tonnes
Cowes	£60,000	10,000–20,000m^3 removed every other year; wharves require dredging every five years.
Fowey	£80,000	Deposition of river–borne sediments
Grampian RC	£180,000	Costs for 13 harbours; Bucke, Burghead and Macduff require dredging to maintain commercial viability.
Hamble	Minimal	Main navigable channel is self scouring.
Harwich	£1.2M	Dredging requirements are extensive.
Heysham	£383,000	Siltation is a major problem, around 0.15m a week.
Holyhead	£17,500	Annual maintenance dredging.
Ipswich	n/s	Deposition as a result of sediment brought down by the R. Orwell necessitates dredging every 5–7 years.
Kings Lynn	£200,000	On–going problems at shifting navigable channels. Major changes occurred in 1943, 1971 and 1980 when the outfall of the Great Ouse changed its position by several miles. Kings Lynn approaches surveyed every two weeks; navigation buoys altered up to 100 times a year.
Liverpool	n/s	1.3M tonnes dredged from approach channels in 1992.
Peterhead Bay	n/s	Dredging undertaken every 3–5 years to keep quay and jetties free; varying amounts removed: 1982 – 1500m^3 1986 – 455m^3 1987 – 3000m^3 1992 – 2175m^3
Poole	n/s	100,000m^3 removed annually.
Rye	£45,000	Siltation a problem due to flood tides and a flashy catchment.
Salcombe	£2,000	Dredging needed every four years.
Sheerness	n/s	Deposition occurs in Chatham dock approaches and close to the Sheerness approaches; maintenance and capital dredging needed.
Stranraer	£10,000	Channel has been dredged to 5m; maintenance required every five years.
Tees & Hartlepool	£1M	Over ten years, 1.55m^3 has needed dredging.
Teignmouth	£100,000	Daily dredging undertaken to keep port open.
Tilbury	n/s	Deposition restricts the navigation to vessels less than 12m draught.
Whitby	n/s	70% of deposition due to river deposition, remainder brought in by tides. 78,000 tonnes/year removed over last three years.
Wisbech	£18,000	R. Nene is generally self scouring, but recent low flows have led to a deposition problem and movement of the navigable channel. Authority has had to remark the channel at a cost of £45,000. Dredging undertaken in the swinging basin and along the quay.

Note: n/s = Not specified

Table 8.2 Summary of dredged material licensed and disposed of at sea.

Country	Year	Licences Issued	Licensed Quantity (tonnes)	Wet Tonnage Deposited
England and Wales	1987	107	28,689,146	38,692,856*
	1988	131	61,645,223	34,691,093
	1989	138	66,408,100	40,810,718
	1990	135	63,983,920	33,728,978
	1991	108	57,782,520	39,886,812
Scotland	1987	35	8,813,850	3,927,264
	1988	25	4,148,690	3,506,685
	1989	27	4,252,950	3,154,756
	1990	21	3,031,960	2,109,114
	1991	26	5,147,245	2,788,611

It is important to recognise a number of inherent conflicts between the various uses of estuaries. By their nature estuaries tend to maximise frictional drag to dissipate tidal energy and, hence, reduce flood risk to adjacent low lying areas. This is achieved by developing a wide, shallow cross-section with extensive mudflats and saltmarshes. Such features are important natural habitats and valued conservation resources. However, port and harbour operations require deep channels and narrow inter-tidal areas. Controlling morphological development, through dredging and channelisation works, has led to a gradual increase in erosion and flood risk within an estuary. Embankments, for example, confine the channel and cause a reduction in the width/depth ratio which in turn leads to an increase in the tidal amplitude and thus flood risk. The normal response of raising embankment height may only lead to a greater tidal amplitude and flood problems (see Chapter 9).

The effects of sea level rise within estuaries will be to produce deeper channels unless sediment accumulation can match the rise. Whilst this may be of benefit to port and harbour operations, it can have significant consequences for conservation and flood defence interests. This is well illustrated by the Humber, where present sea level rise is calculated at 3.5mm a year, and possibly rising to 6–15mm a year by 2030. In order to maintain its present channel form, the estuary would have to accumulate 20Mm^3 a year; this figure would rise to 90Mm^3 a year if sea level rose at 15mm a year (IECS, 1994). Any shortfall in these figures will result in an increase in wave activity in the deeper water, and an increased tidal range which would progress further inland. As described below, current sources of material provide around 7Mm^3 (Table 8.4), emphasising the critical need to take a strategic view for sediment budget management within estuaries.

The Causes of Estuary Sedimentation

Estuaries are links between the river systems and the coast, where the two-way flow of fresh and salt water interacts to produce unique depositional environments. Most estuaries have developed over the last 6000 years or so, after the post glacial rising seas flooded pre-existing deep river valleys or, as in Scotland, glacial troughs. These drowned valleys quickly began to infill with sediment carried by the rivers and from marine sources, developing the complex range of mudflats, saltmarshes, dunes, gravel beaches and subtidal channels that characterise British estuaries.

The rapidity of infill, and consequently the geomorphological stability of the present estuary, depends on a number of factors such as:

- the size and shape of the initial inlet;

- the rate of land subsidence or uplift;

- the availability of sediment;

- the tidal and fresh water inputs;

Table 8.3 Beneficial uses of dredged material from the port of Felixstowe (after Ash, 1994).

1.	Sea Defence bund at West Parkeston for the NRA, as part of a grant-aided capital scheme. (250,000³ gravel).
2.	Developement land for Sea Containers Limited (200,00m³ of sands and gravels).
3.	Low water gravel berms for foreshore stabilisation at Trimley for the NRA (50,000m³ gravel).
4.	Low water berm of rock and clay plus beach replenishment at Fulton Hall Point for the NRA (up to 100,000m³ total).
5.	Foreshore replenishment at Horsey Island for the NRA (100,000m³ sands and gravels).
6.	Low water berm to provide trickle recharge at Naze North for the NRA (50,000m³ sands and gravels).
7.	Construction of rock groynes for the Naze coast protection scheme for Tendring District Council (10,000 tonnes).
8.	Creation of crustacea habitat in a natural deep trench using rock and clay (400,000m³).

- human activity, principally from land reclamation and, more recently, dredging.

Estuaries are the tidal mouths of rivers, but they do not behave as rivers. First, the size and shape of a river is a function of the discharge, which itself is a function of the catchment size and characteristics. This is not true of an estuary in which the tidal discharge is controlled by the size of the channel into which the tidal water moves. Second, rivers tend to develop a shape and pattern which **minimises** the frictional drag of water on the bed thus optimising the transport of water to the sea. The opposite is true of estuaries which tend to **maximise** frictional effects in order that the tidal energy is fully dissipated within the confines of the estuary channel. This means that estuaries are normally extremely shallow and wide – a typical channel such as the Thames may be 5km wide at the mouth and yet have an average depth of only 10m (a width depth ration of 0.002), whereas above Teddington it has a width of 300m and a mean depth of 3m (a width depth ratio of 0.01).

Table 8.4 Estimated sediment budget (Mm³) for the Humber Estuary (after IECS, 1994).

Sediment Source	Sand	Silt	Total
North Sea	?	4	4Mm³
Holderness Cliffs	0.3	0.7	1Mm³
Holderness nearshore	0.7	1.8	2.5Mm³
Ouse/Trent Catchment	?	0.2	0.2Mm³
Total	1Mm³	6.7m³	7.7m³

One of the principle ways in which estuaries dissipate tidal energy is to develop a morphology which rapidly narrow landwards thus forcing the tide into a more hydraulically inefficient channel. The length of the channel and the rate of narrowing depend to a large extent on tidal range. Tidal range is partly determined by the tidal dynamics in the open sea, but more important is the morphology of the estuary itself. Tidal amplitude is increased by the funnelling effect of the channel and, simultaneously, decreased by the frictional drag exerted by the channel. The other important control of estuarine morphology is open sea wave penetration. The ratio between these two energy inputs can account for most of the observed differences between estuaries. Three major tidal types may be distinguished:

(i) **micro-tidal** (0–2m tidal range); processes are dominated by freshwater discharge, with wind-driven waves tending to produce spits and barrier islands which enclose the estuary. The NCC Estuaries Review (Davidson et al, 1991) identifies only six micro-tidal estuaries in Britain, including Christchurch and Poole harbours.

(ii) **meso-tidal** (2–4m); dominated by strong tidal currents, although the limited tidal range means that tidal flow does not extend far upstream. Examples include the estuaries of the East Anglian coast between Yarmouth and Dunwich.

(iii) **macro-tidal** (over 4m); strong tidal currents may extend far inland. Examples include the Thames, the Severn and the Humber.

The sediment budget of an estuary controls both its long-term development and the short term response to human uses. The most important control of sediment movement is the tidal range which determines **tidal currents** and **residual currents** and, hence, the rates and amount of sediment movement (Figure 2.10). This is a very complex subject, but a number of points are worthy or note. Tides are profoundly modified as they enter the shallow, confined estuary waters; amplitude is increased by the drag and by energy concentration in the narrowing channel; the symmetry of the tidal wave is modified by the drag on the estuary bed. This tidal asymmetry can set up a dominant pattern of sediment movement within the estuary:

(i)   **flood dominant estuaries**; relatively low tidal flats and shallow sub-tidal channels lead to an increase in the mean depth at high tide. The high tide wave crest moves faster in its deep water than the shallower wave trough at low tide. Thus, the wave crest can "catch up" with the trough which travels in the shallower water at low tide. This leads to a shorter flood tide and longer ebb tide, giving higher flood velocities;

(ii)   **ebb dominant estuaries**; where an estuary has high inter-tidal flats and deep central channels the average depth taken across the whole channel at high water may actually be less than taken across the channel at low tide. In this case the wave trough moves faster than the wave crest and the ebb tide becomes shorter and faster than the flood.

The effect of tidal asymmetry is to give a net bias to the sediment movement in the estuary. Flood dominance is often associated with a net input of sediment from the sea – the estuary becomes a **sediment sink** – while ebb dominance results in the loss of sediment, usually involving erosion of the estuary bed or banks and the estuary becomes a **sediment source**. Such net movements are, however, mainly of the bed load of the estuary. Net movement of suspended load is associated with the relative length of time during which accretion is able to occur at high tide slack water compared to low tide slack water. If the high tide accretionary period exceeds that at low tide then a net landward movement is imparted to the sediment so that the estuary becomes a sediment sink.

Sedimentation rates in estuaries depend not only on the tidal currents but also on such factors as the availability of sediments, the amount of wave activity in the estuary and a wide range of biological activities including the effects of algae and higher plants. Rates of sedimentation vary enormously between estuaries, between the various zones within an estuary and, perhaps most important, between seasons or even tidal events. Sedimentation in the sub-tidal channels of estuaries can be as high as 150–250mm per week as in the Heysham channel (although this may be partly due to dredging activity). On tidal mudflats average sedimentation rates lie between 1 and 10cm per year, but this reflects a balance between higher short term accretion and subsequent partial re-erosion of the deposited sediment. On tidal saltmarshes accretion rates rarely exceed 1cm, although here re-erosion of deposited sediment is relatively rare and restricted to periods when extreme storm waves can penetrate into an estuary.

Sea level changes are one of the major reasons for channel change in estuaries. The rapid sea level rise following the last (Devensian) glaciation resulted in the re-establishment of the estuaries around the British coast. Most of these estuaries had previously existed during the last inter-glacial period some 150,000 years ago, but their channels had been profoundly changed during the ensuing glacial period when sea levels were 100m lower than at present and fluvial and glacial processes removed most traces of tidal morphology. As sea level rose during the Holocene, so some of these estuaries rapidly developed a new equilibrium morphology. In some cases, however, the valleys cut by rivers in the south of the country, or by glaciers in the north, were so deep as to prevent sedimentary infill and these remain as so called rias (such as the Dart in Devon) or fiords (such as the Cromarty Firth in Scotland). Such estuaries may be regarded as immature or undeveloped.

In the Irish Sea and the southern North Sea abundant glacial sediment and relatively shallow valleys meant that estuaries developed an equilibrium morphology relatively quickly. In such estuaries as the Thames and Blackwater, however, this morphology has recently been subjected to changes which threaten the integrity of their flood defences, as well as their conservation status. These changes are probably due to a combination of sea-level rise and land subsidence and are characterised by erosion of the upper inter-tidal mudflats and recession of existing saltmarsh, is taking place. The increased tidal volume and energy within the estuary is accommodated by an increase in overall width, but a relative decrease in depth so that frictional drag of the tidal inputs is increased.

Where flood embankments have been built, however, the increase in estuarine width has been inhibited. Here the loss of saltmarsh and upper mudflat is not compensated by their landward transgression and thus an overall loss of these important inter-tidal areas is occurring. This process, termed **coastal squeeze**, results not only in the loss of important ecological habitat, but also reduces the efficiency of the estuarine form in controlling tidal inputs. The effect is to increase tidal amplitude and velocities which, in many cases, is causing undermining of the existing flood defences as well as increased flood risks.

There is little known about the supply of sediment to many estuaries, although it is readily appreciated that there are three potential sources: river-borne sediments, coastal erosion and from the sea. In Britain, annual rates of sediment yield from river catchments are generally in the order of 100 tonnes/km^2 or less. Yields are generally higher for upland catchments and rates of over 200 tonnes/km^2 per year have been reported from the Pennines and North Yorkshire Moors. Much of the coarse sediment is, however, "stored" as short-term within channel accumulations (e.g. bars) or long-term floodplain deposits, and does not reach the sea. The exception is the River Spey which carries coarse sediment right to its mouth, where it is dispersed by wave action. Fine sediments are carried into estuaries which act as major depositional sites.

The erosion of coastal cliffs can supply considerable volumes of both coarse and fine sediments to coastal cells. Sea bed erosion can also be an important source of sediment to estuaries. Sediments are primarily transported by littoral drift, waves and tidal currents to depositional areas known as sinks. The SCOPAC Sediment Transport Study, for example, identified a series of principal transport pathways (Figure 2.23; Bray et al, 1991) through which sediment moved to sink areas such as The Solent. This large estuary receives sediment from the erosion of cliffs on the north coast of the Isle of Wight and Christchurch Bay to the west with some sediment supplied from the east. It was noted, however, that the sediment sinks in The Solent had been adversely affected by coast protection works (i.e. reduction of coast erosion and interception of littoral drift) and increased dredging since the 1950s, leading to lowering of banks such as Solent Bank and Hamilton Bank. In the English Channel sea deposits are relatively thin and in some areas are absent all together, whereas north of a line from the Thames to the Severn there is considerable thickness of transportable debris left

there during the ice age (Robinson, 1961). Thus, offshore deposits can be a significant source of material in some estuaries.

The relative importance of these three potential sources will vary according to the estuary. In general, most will receive their sediment from coastal and marine sources, with the exception of some Scottish rivers such as the Spey. The Humber, for example, is one of the few rivers where the sediment budget has been broadly established (Table 8.4). The Holderness coast to the north, eroding at around 1.8m a year, is believed to supply 3.5Mm^3 a year, of which 70% is fine material. Fluvial inputs to the Humber are relatively low, in the order of 0.2Mm^3 per year, although it is considered to have been significantly higher in the past. Indeed, around 1850 it may have been 0.75Mm^3, reflecting the widespread mining activity in the catchment at that time (Pethick, 1993).

The supply of sediment to an estuary can vary over time and can lead to significant changes in the estuary character. In this context, the physical development of the Humber over the last 150 years has been assessed from navigation charts (Pethick, 1993). Since 1850 the estuary has been steadily decreasing in size as mudflats and inter-tidal saltmarshes have formed and the channel has slowly filled with sediment. The average rate of accumulation has been 2Mm^3 a year, or a vertical accretion of 6.5mm a year. However, this disguises major changes in the pattern of sedimentation, with the estimated rates declining from 7Mm^3 a year between 1900-1920 to 1.5Mm^3 by 1970, a decrease of 78% (Figure 8.1). Analysis of the most recent charts suggests that the level of accumulation has fallen, so that in the last decade zero net accumulation has occurred. This decline in sediment may result from:

- a reduction in the background sedimentation from the North Sea;

- a reduction in sediment supplied by erosion of the Holderness Cliffs and nearshore zone as a result of coast protection works;

- interruption of sediment transport into the estuary, as a result of coastal defences and, possibly, dredging at the estuary mouth.

Dredging can result in important modifications to the natural regime. The hydraulic effect of dredging the sub-tidal channel of an estuary is dependent on the relative size of the channel to its

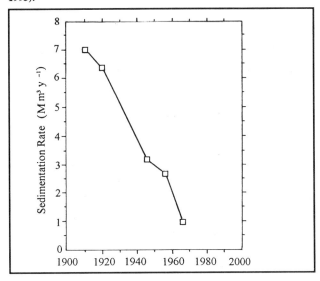

Figure 8.1 Sedimentation rates in the Humber Estuary (after Pethick, 1993).

inter–tidal area. Where a deep navigation channel is dredged in the bed of a formerly wide shallow natural estuary and the effect is to produce **ebb asymmetry** and a tendency towards loss of sediment from the inter–tidal zone will result. In many cases the loss of inter–tidal sediment would be via the sub–tidal channel so that, were this not dredged, sediment would be contained within the system, providing a natural negative feedback mechanism.

In a narrow confined channel the effect of dredging is to produce **flood asymmetry**. In such channels the effect of dredging is to set up a movement of sediment into the deepened channel from marine sources – again resulting in negative feedback. This type of response was recognised by Inglis and Allen (1957) in their classic study of dredging in the Thames estuary. They suggested that the removal of 3Mm³ per year of dredged sediment to the outer estuary merely resulted in its immediate transfer back into the dredged channel. Their suggestion that the dredged spoil should be removed entirely from the estuary or nearshore environment was adopted with the result that dredging was reduced to 0.25Mm³ per year.

Land reclamation for agricultural or industrial use can have a significant effect on estuary morphology and, hence, sedimentation. The enclosure of the River Dee illustrates this point in a dramatic way. Reclamation began around 1730 and continued up to 1986 (Figure 8.2), with much of the land claim used for industry, housing and roads. The gradual reduction in the size of the estuary has been accompanied by expansion of saltmarsh (Doody, 1992); the reduction in tidal volume is believed to have reduced scour and helped accelerate the

natural sedimentation in the estuary. The introduction of Spartina around 1930 probably enhanced this trend, with accretion rates reaching around 25mm a year (Marker, 1967). These changes have been partly responsible for the gradual relocation of the estuary ports closer to the sea. Chester, for example, was a major part up to around 1300. Parkgate was established as a port but was abandoned in the 1820s because of the saltmarsh growth.

It should also be noted that river discharge and velocity can be an important control on the rate and pattern of sedimentation. Modification of the flow regime may, therefore, have significant effects on estuary deposition. For example, abstraction of water from the Welland, Nene and Ouse to reservoirs such as Grantham Water and Empingham reservoir can reduce flows leading to an artificial drought flow below the abstraction points. Reduced river flows may cause deposition in the upper estuary as silt then moves upstream to the tidal limit from the seaward end. At periods of flood, silt will move from the upper estuary to the lower estuary.

The Identification of Vulnerable Areas

Most estuaries have developed equilibrium morphology since the post glacial sea level rise. This adjustment is slow and inexorable, achieved mainly through sedimentation rather than erosion. Some estuaries, however, are less well adjusted due to factors such as lack of sediment (e.g., the Dovey), uplift (e.g., the Dornoch), subsidence (e.g., the Orwell) and the seal level rise. The estuaries of the south west of England, for example, characterised by deep rocky inlets and low levels of sediment inputs from marine sources are still adjusting to the rapid sea level rise of the early Holocene. In contrast, the estuaries of the Irish Sea coast, such as the Dee and Ribble, with relatively shallow initial inlets and high suspended sediment contribution from the sea, have adjusted not only to Holocene sea level changes, but also to subsequent intertidal reclamation. Similarly, on the east coast, the estuaries of the Essex and Suffolk coast, despite rapid subsidence, have received sufficient sediment to allow them to attain a stable morphology which has only been interrupted by extensive reclamation. There is, however, no systematic classification of the sedimentation regime of British estuaries, although the JNCC Estuaries Review (Davidson et al, 1991) identifies the tidal character. Further site details for

107

Figure 8.2 Reclamation in the Dee Estuary (after Doody, 1992).

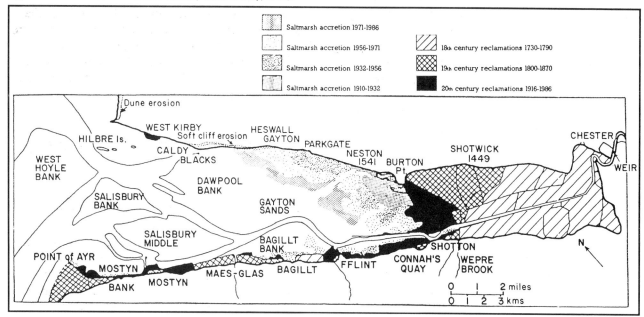

individual estuaries are contained within a 7–volume inventory prepared by JNCC (Buck, 1993).

The Effects of Development

The effects of development on sediment yield in river catchments have already been discussed with reference to hillslope erosion (Chapter 3), flash floods (Chapter 5) and channel instability (Chapter 6). It is unlikely, however, that any of these factors will significantly influence the supply of sediment into estuaries as much will be stored within the river system. This may not have always been the case as the high 19th century sediment yields of the Humber are believed to have been influenced by the dumping of mine waste in the upland areas of its catchment (see above). **Construction of reservoirs** in a catchment may lead to changes in the rate and pattern of deposition within an estuary, as described earlier.

The sediment contributed by the erosion of coastal cliffs can be influenced by development and will be discussed with reference to coastal landsliding in Chapter 10.

Within an estuary, **land reclamation** for development and the construction of **flood embankments** can both lead to modifications within the estuary system as a whole and may lead to changes in the pattern of sedimentation and erosion. However, the most significant effects of these developments are related to changes in flood risk and these will be discussed in Chapter 9.

The Significance for Conservation

Mudflats and saltmarshes support rich habitats for invertebrates and plants which supply food and shelter for other animals (Davidson et al, 1991). A variety of habitats have established in subtidal, intertidal and terrestrial settings containing unique vegetation communities, thousands of invertebrate species (including one of Britain's rarest moths, the Essex Emerald), aquatic communities, many species of fish, amphibians and reptiles. British estuaries are also of major national and international importance for migrant and wintering waterfowl and breeding birds. Many bird species are dependent on the abundant food supply within estuaries. In recognition of their wildlife importance, many estuaries have been designated or otherwise identified under a variety of national and international measures; Ramsar sites, Special Protection Areas (SPAs), SSSIs, National Nature Reserves, Local Nature Reserves (for details see the JNCC Estuaries Review, Davidson et al, 1991).

Mudflats and saltmarshes are dependent on the repeated accumulation and erosion of fine sediments. Mudflats constantly react to storms of varying intensity by steepening or flattening by imperceptible amounts to dissipate the wave energy (i.e. they act like mud beaches). To react in this way the need to accumulate sediments and these sediments must be capable of redistribution throughout the mudflat profile in response to a storm event. Thus, depositional and erosional processes are fundamental to their long term stability.

Mudflat sedimentation is controlled by the velocity and shear stress of the tidal and wave currents, by the concentration of the suspended sediment within the flow and by the size of the suspended sediment grains (or the manner in which they flocculate together, the size affecting the fall velocity of the grains). Thus given low shear stresses, high suspended sediment concentrations and relatively large grained material, sedimentation will be rapid – rates of centimetres of sedimentation have been recorded in a single tide. The average rate of sedimentation for a mudflat, as measured over a period of years, however, is unlikely to be greater than 1cm per year – reflecting a **balance between erosion and deposition** both at a tidal cycle time scale and over longer periods involving storm events of varying return intervals. Sediment deposited on the upper mudflats during a flood tide, for example may be re-entrained on the succeeding ebb and carried down to the low water mark where it may be re-deposited. This process can be repeated over and over again – the sediment recycling between high and low water but very little is either lost from the inter-tidal zone of finally deposited there. In stable or accreting estuaries, the small percentage of the total cycling 'pool' of sediment which is permanently deposited on the mudflats is compensated by a small input from the estuary waters so that a net balance is achieved.

As the mudflats grow in height so they also develop horizontally and eventually assume the form of a wide inter-tidal "beach" which is capable of dissipating the wave energy which is imposed on it. The reduction in wave energy caused by these wide inter-tidal flats allows rapid sedimentation at their upper, shoreward extremity where only infrequent high energy storm waves act to re-entrain the deposited material. As their elevation increases so their tidal inundation period is lowered until they may be covered by tidal water for less than 25% of the 12.4 hour tidal cycle. Under these conditions of low wave energy the long exposure to the air, vegetation may develop on these upper mudflats, forming embryo **saltmarshes**.

The development of a saltmarsh is dependent on a fronting mudflat which can dissipate wave and tidal energy sufficiently to allow sedimentation and vegetation colonisation to proceed. Conversely the marsh can act as a reservoir of sediment, capable of supplying the mudflat during extreme events when an overall widening and flattening of the intertidal zone occurs. The importance of erosion and deposition to estuary management is clear; any

inhibition of sediment movement within the intertidal zone will affect the stability of both the mudflat and saltmarsh forms and, consequently, the habitats which they support. In this context, fixed flood defences can prevent the free adjustment of the saltmarsh and mudflat balance, and, hence, may lead to degradation of these features.

Summary: The Significance for Planning and Development

The **sediment budget** of an estuary can have profound implications for the development potential and the human use. Sediment movement is dependent on the tidal currents which may be asymmetric, i.e. exhibiting a higher velocity on either the **flood** (landward) or ebb (seaward) tides. This asymmetry is to a large extent dependent on the configuration of the estuary channel:

(i)   **flood dominant estuaries**; characterised by low tidal flats and shallow subtidal channels. The high flow velocities during the flood tide can lead to a net input of sediment from the sea, with the estuary acting as a **sediment sink**;

(ii)   **ebb dominant estuaries**; characterised by high tidal flats and deep sub-tidal channels. The high flow velocities during ebb tides can lead to the net erosion of the bed or banks, with the estuary acting as a **sediment source**.

Sedimentation within estuaries has caused major problems for many British ports with periodic dredging required to keep the port facilities open to vessels. Two main types of dredging operations occur: **maintenance dredging**; to keep the navigable channel open and **capital dredging**, for example, of new channels or berths, deepening or changing existing channels to ensure that the port can be competitive by handling larger vessels.

Most port and harbour facilities require dredging. The exceptions are mainly the estuaries where the use is mainly recreational, such as the Artro in West Wales. For example, dredging requirements at both Harwich and on the Tees cost over £1M a year; considerable volumes of material are removed from the estuary bed – 1.3M tonnes a year from the Mersey alone; King's Lynn and Wisbech are affected by shifting channels which have to be regularly resurveyed.

The costs of managing sedimentation problems can be a significant factor in determining the efficiency of a port. Indeed, rising costs and increased competition has led to a trend for the relocation of many port facilities downstream, closer to the river mouth, where the channels are deeper and dredging requirements are less of a constraint. Felixstowe and Tilbury are examples of this pattern. Port authorities may also be faced with considerable difficulties in disposing of the dredged spoil.

Estuary processes are very complex and highly sensitive to the effects of development and use. Whilst the formation of extensive saltmarshes and shallow cross sections help increase the dissipation of tidal energy and, hence, reduce flood risk, such wide, shallow channels are unsuitable for navigation and port developments. Dredging can, however, lead to deeper more erosive channels characterised by the net decrease in saltmarsh and mudflat area, heightening flood problems. The construction of flood embankments can also increase flood risk by tending to confine the channel, leading to a decline in the conservation resource (i.e. intertidal squeeze; Davidson et al, 1991).

The effects of sea level rise in some estuaries will be to produce deeper channels and larger tidal ranges, with heightened flood risk. To minimise these effects, estuaries will need to accumulate sediment at an enhanced rate, producing obvious conflicts with port and harbour interests.

Although channel deposition does not directly raise land use planning and development issues, it is important that the potential problems are taken into account by planners, especially when:

- considering the development of new reservoirs or other structures that could significantly affect river discharges and flows;

- considering the provision of defences on the open coast which may interrupt or interfere with sediment supply to estuaries and, hence, potentially exacerbate the effects of sea level rise;

- identifying potential sites for the disposal of dredged material on land.

Chapter 8: References

Bray M.J., Carter D.J. and Hooke J.M. 1991. Coastal Sediment Transport Study. Reports to SCOPAC. Portsmouth Univ.
Coughlan J. 1979. Aspects of reclamation of Southampton Water. In B. Knights and A.J. Phillips (eds) Estuarine and Coastal Reclamation and Water Storage, 99–124. Saxon House.
Davidson N.C. and others 1991. Nature conservation and estuaries in Great Britain. NCC.
Doody J.P. 1992. The conservation of British saltmarshes. In J.R.L. Allen and K. Pye (eds) Saltmarshes: Morphodynamics, conservation and engineering significance, 80–114. Cambridge University Press.
Inglis C. and Allen F. 1957. The regimen of the Thames estuary as affected by currents, salinities and river flow. Proceedings of the Institution of Civil Engineers, 7, 827–868.
Institute of Estuarine and Coastal Studies 1994. The Humber estuary: coastal processes and conservation. English Nature.
Marker M.E. 1967. The Dee estuary: its progressive silting and saltmarsh development. Transactions of the Institute of British Geographers, 241, 65–71.
Pethick J. 1993. Holderness, The Humber and the North Sea. Paper presented to the KIMO Conference, Hull.
Robinson A.H.W. 1961. The hydrography of Start Bay and its relationship to beach changes at Hallsands. Geographical Journal 127, 63–77.

Chapter 8: Suggested Reading

Pethick J. 1984. An introduction to coastal geomorphology. Arnold Press.

9 The Coast: Flooding of low lying areas

The Nature of the Problems

It has been estimated that over 5% of the population lives in areas below 5m AOD, and, hence, are at risk from coastal flooding. Although many areas are protected by sea defences, coastal lowlands should always be regarded as being at some degree of risk. Low lying coastal land is extremely vulnerable to events which involve either overtopping or breaching of sea defences, especially because of the speed of flooding in such circumstances. Tidal floods can cause extensive damage and distress in the affected area; this is clearly illustrated by the 1990 Towyn floods. On the morning of 26 February sea water overtopped and breached the flood defences along the Clywd coast, inundating 2800 homes in towns and villages from Pensarn to Rhyl. The floodwater was deepest at Towyn where 5000 people were evacuated to neighbouring towns. Although there were no drownings it has been suggested that as many as 50 may have subsequently died from flood-related trauma (Welsh Consumer Council, 1992). However, the floods are particularly memorable for the problems arising from underinsurance and the difficulties encountered during the subsequence clean up, reconstruction and repair.

However, the event was not unique. Britain has a long history of damaging coastal floods, especially in the low lying areas of the east coast and the Thames estuary (Table 9.1). At least seven floods have had major impacts since the 13th century (Jensen, 1953), including the 6 January 1928 floods on the Thames when fourteen drowned in London basements and 4000 were made temporarily homeless. However, the greatest sea flood, in terms of the resulting damage, occurred on 31 January 1953 which inundated over 800km² of eastern England (Figure 9.1). Over 300 died and extensive damage was caused by what was then the highest ever recorded tide levels. Thousands of homes and factories were damaged at an estimated cost

between £900M (Parry and others, 1991); there were 1200 breaches in the sea defences (Steers, 1953). The situation was even worse in the Netherlands where 1600km² were flooded and 1800 lives lost (Volker, 1953).

The 1953 storm surge levels, up to 3m above normal, have since been exceeded, particularly in 1978; however, improvements to the sea defences and flood warning systems, established after the 1953 disaster, ensured that the impact was comparatively small. Even so, thousands of houses were flooded in January 1978 from the Humber to the Thames estuary; at Wisbech an old woman drowned in her home; at King's Lynn there were power failures and 58 children and handicapped people had to be evacuated from a flooded hospital. At Swalecliffe on the north Kent coast nearly 200 houses and 558 caravans were flooded by 0.28m of water, leading to property damage of £1.03M and considerable disruption to life, worry and health effects (Parker 1991). The "intangible" damages arising from the flood at Swalecliffe were estimated at nearly £0.7M.

As part of the feasibility study for a sea defence scheme at Swalecliffe, the Flood Hazard Research Centre, at Middlesex University, identified 230 houses and 588 caravans located in flood prone areas (Parker et al, 1983). Here, the **damage potential** increases with the size of the event, from an estimated cost of £103,000 for the two-year flood to £2.5M for the 500-year flood, of which £1.0M was the estimated "intangibles", (Table 9.2). The rise in the costs partly reflect the increase in area affected and, hence, the number of properties, but also the greater depth of floodwater which can lead to considerable differences in the severity of building damage.

Coastal floods are tidal; the continual movement of water in and out of affected buildings, together with the effect of waves and surges, can cause considerable structural damage over and above that

111

Table 9.1 Severe flood events in the Thames Estuary.

Date	Comment
11.11.1099	Severe flooding of the east coast; a flood of previously unremembered height (the Anglo Saxon Chronicles).
1158	Abnormal tides caused extensive flooding on the north Kent coast.
1235	Chatham, Rochester, Gillingham inundated.
1236	Great damage caused by severe tides; considerable loss of life (Matthew Paris, the St Albans Chronicler).
1237	North Kent marshes inundated.
1242	Major flooding on the Thames.
1287	Storm floods on the Kent coast.
1309	Serious storm and flooding of the Thames and north Kent coast.
1332	Tidal inundation at Chatham and Rochester.
1362	The "Grote Mandrenke"; 11,000–30,000 drowned on the coast of northern Europe. Flooding across southern England.
1409	Extensive flooding after severe frost with deep snow – a river flood (?)
1534	River frozen; flooding on thaw.
15.10.1570	Great storm on the east coast; catastrophic damage. Followed by the "All Saints Flood" of 12.11.1570 which left 100,000–400,000 drowned on the continent.
1613	"Dreadful inundation by the sea"; severe flooding.
7.12.1663	Great tide flooded London and the surrounding areas, "There was last night the greatest tide that ever was remembered in England to have been in this river, all Whitehall having been drowned"; (Diary of Samuel Pepys).
1682	Exceptionally high tides at Sheerness and Chatham; severe flooding.
29.1.1701	East coast was flooded.
7.12.1703	Reported to have been the severest storm on record; 8000 dead across Britain. Storm surge caused extensive flooding.
1707	Storms and sea floods.
24.12.1717	Flooding along east coast; one of the greatest recorded storm disasters on continental Europe, about 11,000 dead.
22.2.1735	Severe floods; highest tide on the Thames for 50 years.
5.1.1736	Great tide on east coast; Westminster Hall flooded to a depth of 0.6m.
9.2.1762	Flooding on the Thames
1.3.1791	Very high tide, most cellars in Whitehall full of water; damage to cornlands by the Thames estimated at £20,000. Westminster Hall flooded (4.5m AOD). Extensive damage on north Kent coast.
1825	Exceptionally high tides at Chatham.
1834	Exceptionally high tide recorded.
18.10.1841	Thames flooded.
11.12.1845	Chatham docks flooded.
29.1.1850	Severe flooding.
1852	Exceptionally high tide reported.
20.2.1854	Chatham docks flooded.

Table 9.1 (cont ...)

Date	Comment
1856	Severe east coast floods.
1874	Thames flooded.
1875	Exceptionally high tide reported.
18.1.1881	Exceptionally high tide; Thames flooded.
19.2.1882	Floods recorded.
28.10.1882	Floods recorded.
12.3.1883	Chatham town and docks flooded.
29.11.1897	One of the most destructive tides on record, extensive damage from Norfolk to Dover; Sheerness pier washed away, severe damage to flood defences, buildings flooded to several feet. Its highest level (at Blakeney) was 1.3m below the 1953 storm surge level.
1901	Severe spring tides caused widespread flooding.
30.12.1904	Severe flooding at Chatham (tide recorded at 7.3m).
1906	Flooding of the Thames marshes, sea defences breached and extensive damage to arable land.
1907	Sea defences carried away by high tides and gales.
6.1.1928	Serious flooding in inner London as tides rose above the embankments; 14 drowned. Northerly gales coupled with spring tides created a tide 2m above predicted.
2.2.1935	Force 10 gales, severe flooding.
9.2.1935	Sea walls breached and tidal damage in north Kent.
12.2.1938	Severe flooding at Chatham and north Kent.
1.3.1949	Storm surge led to some breaching on the Medway.
29.11.1951	Tidal flooding along the north Kent coast.
31.1.1953	Severe storm and tidal surge; the East Coast Floods with over 300 dead. Worst floods in recorded history.
12.1.1978	Serious flooding on the east coast, parts of Sheerness flooded to over 1m.

which can be expected for river floods. Inundation by sea water can result in a 10–20% increase in damage due to the effects of saltwater (Cole and Penning–Rowsell, 1981; Penning–Rowsell et al, 1992) when compared with freshwater flooding. The most important effects are:

- impregnation of floor timbers, leading to an increased tendency for dry rot;

- greater damage to salt affected floor coverings;

- rusting and damage to central heating systems;

- increased damage to personal effects and domestic appliances.

A frequent problem associated with coastal areas is the use of flood risk areas for caravan and camping sites, often with severe access restrictions which can make dissemination of flood warnings difficult. Holiday–makers or occupants may face additional risks because of the instability of many caravans, which can be swept away or turned–over by high velocity floodwaters. Further difficulties can arise because of the trend for coastal sites being used for retirement homes by the elderly, who may be less able to respond quickly to warnings of imminent flooding.

The indirect or secondary economic effects of coastal floods have been documented by Penning–Rowsell and Parker (1980) in a study of the problems affecting the Isle of Portland to Weymouth causeway. They demonstrated that the

113

Figure 9.1 The area inundated by the 1953 east coast floods (after Perry, 1981).

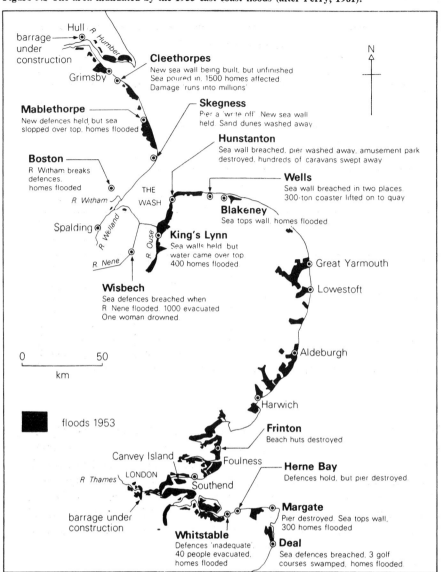

impacts were related to the type of flooding that flooding that occurs in this part of Dorset:

(i)   **percolation floods**; when water seeps through Chesil Beach;

(ii)   **storm surge floods**; overtopping of the sea wall and parts of the beach during high tides and gales;

(iii)   **ocean swell floods**; overtopping caused by very large waves.

The associated return periods and the average duration of traffic and business disruption for the three types of event are shown in Table 9.3a. From this it is clear that flooding can lead to long periods of traffic and business disruption. The actual pattern of disruption is complex, but may involve a wide variety of causeway users (Table 9.3b):

(i)   **disruption to the manufacturing and service sector**; mainly resulting from employees being prevented from travelling to work across the causeway and includes the indirect losses to the Royal Naval establishment on the Isle of Portland (Table 9.3c);

(ii)   **general road traffic disruption**; including goods deliveries, buses and tourist or leisure uses;

(iii)   **mail delays**; indirect costs to the Post Office of £124 were estimated, covering overtime for clearing the backlog of mail caused by floods in 1978 and 1979;

Table 9.2 Estimated sea flooding damage potential at Swalecliffe, Kent (1983 £ values, overtopping of existing sea defences, after Parker, 1991).

	Flood Return Period (Years)			
Properties Affected and Value	2	32	114	500
Residential Damage (£)	1 147	143 163121	193 323061	230 541673
Retail Damage (£)	5 3713	9 16794	9 19580	9 27648
Offices Damage (£)	1 2529	1 8342	1 8342	1 8485
Caravans Damage (£)	143 97253	530 574321	568 665507	588 736657
Industrial Damage (£)	0 0	3 0	5 21363	8 186540
Total Damage (£)	150 103642	686 762578	776 1037853	836 1501003
Intangibles (£)	0	331728	691100	988273

(iv) **emergency services**; during the 1978/79 floods Wessex Water Authority established an emergency centre and the Weymouth and Portland Borough Council organised the emergency works. Both deployed staff in clean−up operations, reconstructing Chesil Beach and reopening the causeway.

This example from the Isle of Portland demonstrates that where there is a high dependence upon a communication link, the effects of flooding can lead to large indirect losses spreading well beyond the flooded area. This is highlighted by the high proportion of indirect to total flood losses, which is estimated to be as high as 93% for percolation type floods and 28−32% for the other types (Table 9.3b). Penning−Rowsell and Chatterton (1977) identified a similar pattern for floods at Pulborough, Sussex and Ashton Vale, Bristol; in both areas indirect damages exceeded 20% of total flood losses.

In many instances, the speed and unexpected nature of coastal flooding can lead to many serious health problems in the affected communities. For example, the tidal flooding at Uphill, near Weston−Super−Mare, on 13 December 1981 greatly affected the health of many residents (Green et al, 1985). The flood followed failure of the sea defences in the early evening. Very few people received any warning, as attempts by the police to raise the alarm were thwarted by the rapidly rising floodwaters. The force of the water broke down doors and some houses were flooded to several feet. The worst affected were those trapped in bungalows. Rescuers reached most areas between 11 p.m. and midnight, 2−3 hours after the flood, with many evacuated to an emergency centre or their relatives' homes.

The population of Uphill was mainly elderly and, hence, particularly vulnerable to the effects of flooding. Several houses suffered over £10,000 of building fabric damage, but the direct losses were probably outweighed by the disruption, worry, loss of health and depression that followed the clean−up operations. Indeed, the elderly are very sensitive to personal losses and may experience a strong feeling of deprivation due to the disruption or loss of objects with sentimental value (Edwards, 1976).

Many of the impacts described in this section have focused on the damages and losses incurred by individual communities, as these are readily identifiable. However, many coastal floods can have an effect over long stretches of coastline, imposing considerable damages to all sectors of the economy, disrupting communications and leaving thousands temporarily homeless. The cumulative damages of many of the more severe coastal floods, such as the 1978 event, are likely to be over £50M and may exceed £900M for disasters

Table 9.3(a) Characteristics of the Main Types of Floods Affecting Chiswell and the Weymouth–Portland Causeway Road (after Penning–Rowsell and Parker, 1987).

	Percolation–Type Flood	Storm–Surge–Type Flood	Ocean–Swell–Type Flood
Estimated return period (years)	0.5	5	50
Depth of flooding in Victoria Square, Chiswell (metres a.o.d.)	3.70	6.50	6.50
Average duration of causeway traffic disruption (days)	0.5	1.0	2.0
Average length of business disruption in Chiswell (weeks)	0.5	2.0	3.0

Table 9.3(b) Summary of the Best Estimate of Indirect Costs and Losses Per Flood from Disruption by Flooding of the Weymouth to Portland Causeway Road (after Penning–Rowsell and Parker, 1987).

	Percolation–Type Flood (£)	Storm–Surge–Type Flood (£)	Ocean Swell–Type Flood (£)
Manufacturing and service sector disruption	38452	76904	153808
Cost of other road traffic disruption	2497	4993	9986
Reported cost of mail delays	–	124	124
Report costs of emergency services	48	35509	65024
Total estimated indirect losses	40997	117530	228942
Total estimated direct flood damages and indirect losses	44031	415429	707391

Table 9.3(c) Estimates of Indirect Cost of Flooding to Royal Navy Facilities at Portland (after Penning–Rowsell and Parker, 1987).

	Estimated Cost of Wages Lost (£)		Cost of Emergency Boat Service (£)	Total Cost (£)
	Service Personnel	Civilian Personnel		
Percolation flood	2559	12554	48	15161
Storm surge flood	5118	25107	115	30340
Ocean swell flood	10236	50214	230	60680

such as the 1953 floods (Parry and others, 1991).

Coastal flooding is frequently associated with widespread foreshore erosion and offshore accretion as beach or mudflat sediment are redistributed in the face of very high wave energies. Marine erosion can rapidly remove areas of high ground protecting lower lying areas behind. Although permanent changes can result, much of the "damage" to foreshore systems sustained by these events can be restored over the forthcoming months and years, as sediment is transported back onshore by wave and tidal energy. The 1953 storm surge, for example, caused extensive erosion; sand and shingle were stripped from the beaches, dunes were undercut and marshes flooded. After the storm, however, there was considerable accretion as offshore bars, which developed as a result of the storm, moved inshore. Some areas quickly gained more material than they had lost (Barnes and King, 1953).

Amongst the most notable coastal changes associated with extreme storm and flood events are those which occurred on the Grampian coast. At Rattray, for example, around 1700 Stathbeg Loch was formed when the sea inundated the coastal plain and sand dune movement blocked the mouth of the harbour (Lamb 1991). Spurn Head, a spit on the Humberside coast, has been breached on many

116

occasions; in December 1849, for example, a storm caused a 400m wide breach across the neck of the peninsula connecting it to the mainland. The spit was reduced to a series of islands at high tide and the ramparts defending the lighthouse were washed away (De Boer, 1968); the spit has since re-formed, highlighting the cyclic nature of some coastal changes.

The Causes of Coastal Floods

In contrast to floods on hillslopes and floodplains, coastal floods are not the result of heavy or prolonged rainfall, although they too are related to climatic events and can be intensified by large river discharges. On the coast the interactions between storms and tides are critical, producing water levels that may be considerably in excess of the tide levels normally associated with a particular stretch of coast. This relationship between severe weather and coastal flooding is highlighted by the marked seasonal variation of recorded flood events, with the majority of events occurring between November and March.

Flooding of low-lying coastal areas usually occurs during extreme climatic events such as severe storms or gales. However, the factors which influence the pattern and extent of flooding are complex, involving short-term changes in:

● water level
● "natural" coastal defences

The tidal range experienced around the British coast (Figure 2.10) can be severely modified by direct and indirect meteorological influences, leading to the development of **surges** when the actual water height can be considerably in excess of the expected level. These surges can comprise a number of components, including the effect of wind and tidal response, to changes in barometric pressure (water level will rise by 0.01m for every millibar fall in pressure). Although the effect in the open sea is negligible, in shallow coastal waters the shoaling transformation of these waves can cause a surge with a height of 2m or more. Figure 9.2 demonstrates this effect for a storm in the Irish Sea which generated a surge of 2m in the Solway Firth and compensating lower water levels on the east cost of Ireland (these low levels, or negative surges, can present a navigational hazard to shipping).

The severity of a surge is related to the size and track of a storm and its proximity to the coastline, together with the wind direction and fetch. Bretschneider (1967) has demonstrated that maximum water level rise along the coast resulting from the effect of wind was achieved at low windspeeds when the wind is parallel to the coast and at very high speeds when the wind is normal to the coast. Fetch, in association with windspeed, determines the amount of water piling up at the coast and the wave energy associated with the storm.

The North Sea is particularly susceptible to storm surges entering from the North Atlantic or generated by storms crossing the sea itself. There has been a catalogue of damaging surge events, most notably 1825, 1894, 1897, 1906, 1916, 1921, 1928, 1936, 1942, 1943, 1949, 1953, 1969, 1976 and 1978. All appear to have been associated with strong north west winds accompanying the passage of deep depressions moving across the sea to the Norwegian coast. The 1953 storm was the most memorable event in Britain and has been extensively documented (e.g. Robinson, 1953; Rossiter, 1954; Steers 1953).

The sequence of events began with a depression forming to the south west of Iceland on 29 January which began to move towards the North Sea. The position of its centre at six hourly intervals from 0000 hours on 30 January is shown on Figure 9.3(a); the pressure falling from 996 mbar to 968 mbar by midday of the 31st. The depression continued to move south and east, with the gale force winds veering to the north west, increasing the fetch. This encouraged a southward flow of water along the east coast, as the wind drift was deflected to the right by the earth's rotation. The whole of the southern North Sea was raised, by more than 2m south of the Humber and over 3m on the Netherlands coast (Figure 9.4b). After midnight, on 1 February, the peak of the surge reached low-lying areas of the east coast, moving progressively southwards (Figure 9.3c) and causing extensive flooding (Figure 9.1).

Many lowland floods arise because of the combined effects of large river discharges and high tides within estuaries. Fortunately the 1953 surge did not coincide with flooded rivers, otherwise the damage could have been considerably more extensive. Such **estuarine floods** occur when the seaward flow of freshwater is impeded by the rising tide. Particular problems can arise with the higher Spring tides which arrive 2-3 days after the full moon. On the Humber, for example, the

117

Figure 9.2 Storm surge in the Irish Sea, January 1975 (after Carter, 1988).

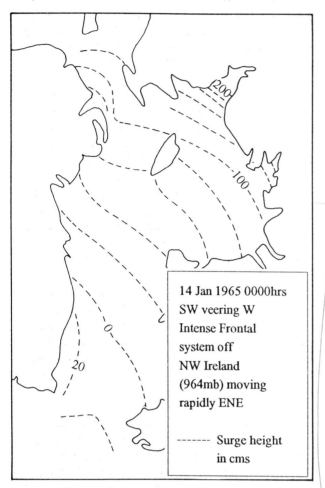

14 Jan 1965 0000hrs
SW veering W
Intense Frontal
system off
NW Ireland
(964mb) moving
rapidly ENE

------ Surge height
in cms

The causes of estuarine floods can be very complex, often reflecting a combination of circumstances. For example, very high Spring tide levels can so reduce the storage capacity of an estuary channel that even a "normal" freshwater flow during the flood tide can lead to the remaining capacity being exceeded. Conversley, very large river discharges can result in flooding even on average tides. A striking example of the combined effects of large freshwater discharge and high tides intensified by a storm surge occurred in the Thames estuary in 1928 (Table 9.1), when the river rose nearly 2m above the predicted level and led to extensive flooding.

Many coastal landforms offer a degree of protection against coastal flooding. **Sand dunes** serve as a natural barrier against high water levels and form effective coastal defences for many communities around the coast. **Beaches** and **shingle ridges** absorb as much as 90% of the wave energy arriving at the coast by continuously adjusting their form (Brampton, 1992a), providing an important component of sea defences either alone or where they front embankments and sea walls. **Saltmarshes** and **mudflats** are also effective in dissipating wave energy. This is clearly demonstrated in Figure 9.4 which shows how waves travelling across a saltmarsh towards a flood defence embankment are reduced by breaking and shoaling. However, a decrease in saltmarsh width or surface elevation will cause an increase in overtopping or require an increase in the crest height of the wall to provide the same standard of protection (Brampton, 1992b). An example of the efficiency of saltmarshes in floodwave attenuation can be found on the Gwent Levels, where there are two types of defence. Where defences are fronted

freshwater input can have an significant effect on high water (Table 9.4), although the importance declines towards the estuary mouth. At Gainsborough a major river flood may increase both high water and low water over 5m.

Table 9.4 Effects of various freshwater inputs upon tidal levels in the Humber Estuary.

Location	Tidal State	Freshwater Input (cubic metres per second)			
		No Input	285 cumecs	570 cumecs	855 cumecs
Keadby	H.W	+3.81m	+4.12m	+4.21m	+4.25m
	L.W.	−0.64m	+0.18m	+0.70m	+1.25m
Owston Ferry	H.W.	+3.69m	+4.45m	+4.63m	+4.75m
	L.W.	+0.18m	+1.83m	+2.71m	+3.48m
Gainsborough	H.W.	+3.38m	+4.82m	+5.09m	+5.73m
	L.W.	+0.5m	+2.98m	+4.55m	+5.67m

Figure 9.3 The 1953 storm surge: the track of the depression (top), the surge levels on 1 Feb in metres (mid) & surge graphs for east coast locations (bot) (after Ward, 1978)

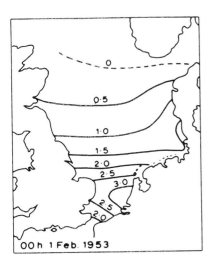

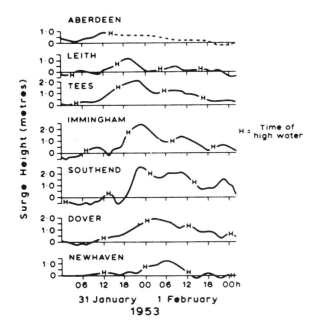

by healthy saltmarsh they comprise an earth embankment with crest heights around 8.8–9.5mAOD; where there is no marsh, the defence levels are at 10.5mAOD and involve a near vertical wall capped by a wave return wall and fronted by an 8m high rock slope (Green, 1984).

Many coastal landforms are very vulnerable to erosion; this can have serious implications for flood risk and coastal defence expenditure. Wakelin (1989), for example, describes how the beaches of East Anglia have lowered and steepened over recent decades (Table 9.5). This has lead to a general decrease in the beach volume in front of sea defences and problems of undermining. Pye and French (1993) have undertaken a review of the saltmarsh erosion and accretion to provide an overview pattern and to evaluate the implication for flood defence management. They have demonstrated that marshes in north west England, Wales, eastern England and much of Scotland show both vertical and lateral accretion, whereas marshes in south east southern and south west England show a predominant trend of lateral and internal erosion; this correspondences to the pattern of crustal rise due to isostatic readjustment.

In the Medway estuary, for example, erosion appears to have begun around 1700, probably in response to natural conditions but exacerbated by mud digging for local brick and cement industries (Kirby, 1990). Erosion accelerated in the early 20th century, partly due to the effects of severe storm surges in 1897 and 1901, which caused breaches in the sea walls around reclaimed marshland, and their subsequent abandonment. The breaching, together with continued mud digging, sea level rise and changes to the regional tidal regime may have led to an increase in current velocities, reinforcing the trend of erosion on the sides of the main channels (Pye, 1993). Mudflats have since lowered, causing marsh creeks to deepen and widen by bank collapse. Higher sea levels have increased the wave attack on the mudflats and marsh edges.

The factors contributing to erosion and accretion of these coastal landforms are complex. Human activity can, however, have a significant effect, most notably through:

- the disruption of sediment **supply** by the protection of eroding cliffs;

- the disruption of sediment **transport** by coastal works such as groynes, breakwaters and harbours (see Chapter 10 for discussion of the effect of the construction of

119

Figure 9.4 The role of mudflats and saltmarshes in flood wave attenuation (after Brampton, 1992b).

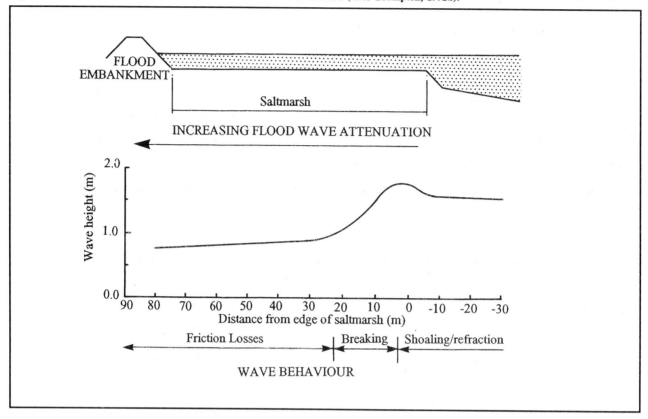

Folkestone harbour).

● removal of sediment from sand dunes, beaches and shingle ridges for use in construction (see Chapter 11).

Other potentially damaging activities include **land claim** which can reduce the extent of saltmarsh and mudflats; **aggregate dredging** may affect the supply of sediment to beaches from offshore sources (the evidence is, however, equivocal), **sand mining** can lead to problems within sand dune areas.

The Identification of Vulnerable Areas

Low lying areas below 5m AOD have been identified on the accompanying 1:625,000 scale thematic maps as areas at risk from coastal flooding (the 5m contour is the first contour mapped by the Ordnance Survey above 0m AOD). However, the **potential** for storm surges and, hence, flooding varies around the British coast (Pratt, 1993):

(i) **the east cost of England**; this coast experiences an average of 19 surges with elevations of over 0.6m above "normal"

each winter. These surges have reached a maximum recorded level of around 3.5m in the Thames estuary in 1921 and 1953.

(ii) **the west coast**; surges are generally associated with strong south–westerly winds ahead of depressions approaching the west of Ireland. It has been found that the number of surge events per year is similar to that on the east coast.

(iii) **the south coast**; this coast is less seriously affected by storm surges. Surge levels exceeding 1m are rare, although wave action is particularly significant in causing flood events (see the Chesil example described earlier).

(iv) **Scotland**; surge levels on the east coast rarely exceed 1m, but can cause problems when combined with periods of extreme river discharges as has occurred in the Moray Firth.

On the south west coast of Scotland, surge levels have been considerably amplified by strong winds blowing straight along the longer sea lochs. A major event occurred on 5 January 1991 when the observed level at Greenock was over 1.5m above the normal tidal level.

Table 9.5 Beach level changes on the East Anglian coast (after Wakelin, 1989).

(a)   Net Beach Level Changes September 1959 to April 1985 (m/yr)

Site	Distance Offshore (m)						
	0	20	40	60	80	100	120
Mablethorpe	−0.03	−0.03	−0.03	−0.02	−0.02	−0.02	−0.01
Trusthorpe	−0.02	−0.02	−0.02	−0.02	−0.01	−0.01	−0.01
Anderby Creek	+0.03	+0.03	+0.03	+0.02	0	0	+0.01

(b)   Average Beach Level Variations (m/yr) Over 0–100m Width of Beach.

Site	Beach Level Variations (m/yr)		
Mablethorpe	0 (1962–74)	−0.10 (1974–80)	+0.04 (1980–83)
Trusthorpe	−0.01 (1962–74)	−0.07 (1974–80)	−0.07 (1980–83)
Anderby Creek	0 (1965–78)	+0.10 (1977–83)	

The Effect of Development

As described earlier many "natural" coastal defences are sensitive to the supply and transport of sediment; disruption to established transport pathways can lead to degradation of these landforms and an increase in flood risk. Coast protection works, sea defences, mineral extraction from dunes or beaches, harbour works and marina breakwaters can all have a significant effect on coastlines where longshore sediment transport is important.

In estuaries, the progressive constriction of the channel through the construction of embankments can lead to a reduction in the storage capacity of the estuarine lowlands and marshes (see Chapter 8). Horner (1978), for example, has demonstrated that high tides on the Thames at London Bridge have risen dramatically over the last 150 years (Figure 9.5), this can only be partly explained by rising sea levels and settlement in the London area caused by groundwater abstraction. The reduction in storage can act to pass the peak of the flood tide upstream, increasing the potential flood problems above the reaches that have been defended, i.e. the opposite of the effects of flood defence on rivers (Chapter 7).

The Significance for Conservation

Flooding is essential to maintain a regular supply of fine sediments to mudflats and saltmarshes (see Chapter 8). However, saltmarsh plants can only tolerate limited submergence; as waterlogging increases, anaerobic conditions will prevail. Sea level rise is, therefore, a major concern for conservationists and shoreline managers as it could result in a reduction in the area of saltmarsh with adverse impacts on those habitats and coastal defences. These concerns have led to considerable interest in the possibilities of **managed retreat** to restore or create desirable habitats and provide improvements to sea defences on low lying coastlines (Posford Duvivier Environment, 1991). This can involve the partial or total abandonment of the existing line of defences where they may be too expensive to maintain or improve to cope with sea level rise. **Deliberately reduced standards of defence** or **tiered defence systems** combine a level of flood protection with the potential for regeneration of valuable saltmarsh and inter tidal habitats (Brooke, 1993). Such options are, of course, only likely to be applicable in undeveloped areas where retreat would not increase the risk of flooding to coastal settlements.

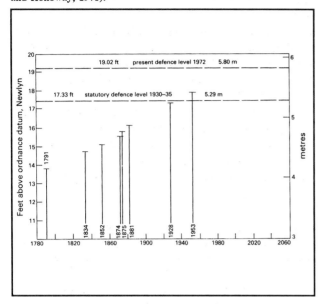

Summary: The Significance for Planning and Development

Tidal flooding of coastal lowlands can be extremely devastating, affecting large stretches of a coastline in a single event. The 1953 floods, for example, caused an estimated £900M of damage and over 300 deaths, inundating over 800km² of eastern England. Flood events can involve overtopping or breaching of defences; the latter can be particularly damaging because of the speed of flooding in such circumstances, and might involve loss of life as well as destruction of property.

Flood events are often associated with widespread coastal erosion and deposition as beaches and mudflat sediments are redistributed by very high wave energies. Although permanent changes may arise, much of the foreshore "damage" may be restored over the forthcoming months and years, as sediment is transported back onshore by waves and tides. It is important to stress, however, that coastal erosion during severe storms can rapidly remove parts of beaches or sand dunes which provide protection for the lower lying land behind; the risk of flooding on the coast can, therefore, change very quickly as natural coastal defences are modified by shoreline processes.

As for lowland river floods, the severity of damage is generally related to the depth and duration of flooding, together with the degree of forewarning. Inundation by salt water can, however, result in a 10–20% increase in damage over and above that which could be expected for river floods. In many instances the speed and unexpected nature of coastal flooding can lead to serious health problems in the affected communities.

Tidal floods are generally caused by storm surges, where low pressure atmospheric conditions and high wide speeds can lead to tides in excess of 2m above the expected tidal level. There is a marked seasonality for coastal flooding with the majority of events occurring between November and March. Although surges may occur around the entire coastline, the North Sea is particularly susceptible with numerous events recorded over the last 500 years, most of which appear to have been associated with strong north west winds accompanying the eastward passage of deep depressions to the Norwegian coast. In estuaries, floods can arise because of the combined effects of large river discharges and high tides.

Although climatic factors determine the occurrence of storm surges, their impact on the coast can frequently be intensified by human activity. Of particular significance is the starvation of sediment to natural coastal defences such as sand dunes, beaches and shingle ridges through coastal defence works and mineral extraction. Encroachment of development, including caravan and camping sites, into low lying coastal areas is also a significant factor in increasing the potential damages associated with tidal flooding, as was described for lowland flooding in Chapter 7. In this context, it is important to stress that low lying coastal lands should always be regarded as being at some degree of risk from flooding because of the dynamic nature of coastal processes and the occurrence of extreme, but rare, storm surges.

From a conservation perspective, regular flooding is essential for supplying fine sediments to saltmarshes and mudflats. Indeed, tidal defence works can result in significant environmental effects elsewhere because of the disruption of this supply of sediment and the "inter-tidal (see Chapter 8).

Coastal flooding should be viewed as a major constraint to development because of the possible loss of life, destruction to property and expense of protecting vulnerable communities. Local planning authorities can ensure that the effects of coastal flooding are minimised by:

- guiding development away from low lying coastal areas at risk from flooding;

- restricting forms of development that could increase the risk of flooding elsewhere, such as mineral extraction from the foreshore and sand dunes or foreshore structures which disrupt the natural movement of sediment along the coast;

- ensuring that coastal land allocated for development can be satisfactorily protected without affecting nature conservation or other environmental interests.

In England and Wales, guidance to local planning authorities and others is provided in:

- Circular 30/92 Development and Flood Risk (DoE 1992a);

- PPG 20 Coastal Planning (DoE 1992b).

No comparable advice has been issued by the Scottish Office (NB. draft advice was issued in March 1995).

Chapter 9: References

Barnes F.A. and King C.A.M. 1953. The Lincolnshire coastline and the 1953 storm-flood.Geography 38, 141–160.

Brampton A. 1992a. Beaches – the natural way to coastal defence. In M.G. Barrett (ed) Coastal zone planning and management, 221–229. Thomas Telford.

Brampton A. 1992b. Engineering significance of British saltmarshes. In J.R.L. Allen and K. Pye (eds) Saltmarshes: morphodynamics, conservation and engineering significance, 115–122, Cambridge University Press.

Bretschneider C.L. 1967. Storm surges. Adv. Hydroscience 4, 341–418.

Brooke J. 1993. Coastal defence and managed retreat. WWF Update 10.

Carter R.W.G. 1988. Coastal environments. Academic Press.

Cole G. and Penning-Rowsell E.C. 1981. The place of economic evaluation in determining the scale of flood alleviation works. In Flood Studies Report – 5 years on. Institution of Civil Engineers, 143–151.

Department of the Environment 1992a. Circular 30/92. Development and Flood Risk. HMSO.

Department of the Environment 1992b. PPG20. Coastal Planning. HMSO.

De Boer G. 1968. A history of the Spurn Lighthouses. East Yorkshire Local History Society. York.

Edwards J.G. 1976. Psychiatric aspects of civilian diasters. British Medical Journal, 17, 944–947.

Green C. 1984. Saltings and sea defence on the Gwent Levels. MAFF Conference of River Engineers, Cranfield.

Green C.H., Emery P.J., Penning-Rowsell E.D. and Parker D.J. 1985. The health effects of flooding : a survey at Uphill, Avon. Flood Hazard Research Centre, Middlesex University.

Groen P. and Groves G.W. 1962. Surges. In M.N. Hill (ed) The Sea, 611–646, Interscience, New York.

Horner R.W. 1979. The Thames Barrier project. Geographical Journal 154, 242–253.

Jensen H.A.P. 1953. Tidal inundations past and present. Weather, 8, 85–89, 108–112.

Kirby R. 1990. The sediment budget of the erosional intertidal zone of the Medway estuary, Kent. Proceedings of Geological Association 101, 63–77.

Lamb H.H. 1991. Historic storms of the North Sea, British Isles and Northwest Europe. Cambridge University Press.

Parker D.J. 1991. Flood disasters in Britain : Lessons from flood hazard research. Flood Hazard Research Centre, Middlesex University

Parker D.J., Green C.H. and Thompson P.M. 1987. Urban flood protection benefits: a project appraisal guide. Gower Technical Press.

Parker D.J., Penning-Rowsell E.C. and Green C.H. 1983. Swalecliffe coast protection proposals: evaluation of potential benefits. Flood Hazard Research Centre, Middlesex University.

Parry M.L. and others 1991. The potential effects of climate change in the UK. HMSO.

Penning-Rowsell E.C. and Chatterton J.B. 1977. The benefits of flood alleviation: a manual of assessment techniques. Saxon House/Gower Press.

Penning-Rowsell E.C. and Chatterton J.B. 1980. Assessing the benefits of flood alleviation and land drainage. Proceedings of the Institution of Civil Engineers 69, 295–315, 1051–1054.

Penning-Rowsell E.C. and Parker D.J. 1980. Chesil Sea Defence Scheme : Benefit Assessment. Flood Hazard Research Centre, Middlesex University

Penning-Rowsell E.C. and Parker D.J. 1987. The indirect effects of floods and benefits of flood alleviation: evaluating the Chesil Sea Defence Scheme. Applied Geography, 7, 263–288.

Penning-Rowsell E.C., Green C.H., Thompson P.M., Coker A.M., Tunstall S.M., Richards C. and Parker D.J. 1992. The economics of coastal management : a manual of benefits assessment tech niques. Belhaven Press.

Posford Duvivier Environment 1991.
Environmental opportunities in low lying coastal areas under a scenario of climatic change. Report to NRA, DoE, NCC and Countryside Commission.

Pratt I. 1993. Operation of the Storm Tide Warning Service. Proc. MAFF Conference of River and Coastal Engineers, Loughborough.

Pratt S and Holloway J.M. 1978. Thames Barrier project : Provision of Services. Building Construction Forum, Feb. 1978.

Pye K. 1993. Saltmarsh erosion and accretion. Proc. MAFF Conference of River and Coastal Engineer, Loughborough.

Pye K. and French P.W. 1993. Saltmarsh erosion and accretion in Great Britain. Report to MAFF. CERC Limited.

Robinson A.H.W. 1953. The storm surge of 31 January – 1 February 1953. Geography 38, 134–141.

Rossiter J.R. 1954. The great tidal surge of 1953. The Listener, 8 July, 55–56.

Steers J.A. 1953. The east coast floods January 31 – 1 February 1953. Geographical Journal 119, 280–298.

Volker M. 1953. La Maree de tempete due ler feurier 1953. La Houille Blanche 2, 207–216.

Wakelin M.J. 1989. The deterioration of a coastline. In Coastal Management, 135–152, Thomas Telford.

Welsh Consumer Council 1992. In Deep Water. Welsh Consumer Council.

Chapter 9: Suggested Reading

Carter R.W.G. 1988. Coastal environments. Academic Press.

King C.A.M. 1972. Beaches and coasts. Arnold Press.

Perry A H 1981. Environmental hazards in the British Isles. Allen and Unwin.

Pethick J. 1984. An introduction to coastal geomorphology. Arnold Press.

Ward R.C. 1978. Floods : a geographical perspective. MacMillian Press.

10 Coastal Cliffs: Landslides and Cliff Recession

The Nature of the Problems

Cliff recession should be viewed as a 3-stage process comprising **detachment** of material (ranging from individual soil particles to enormous coherent blocks of material in landslides), the removal of debris and **transport** by water, and the **deposition** of sediment elsewhere (Figure 10.1). The rate of removal of debris from the foreshore is of particular importance as this material provides protection against further detachment.

Although sub-aerial processes can be important locally, most erosion of cliffed coastlines is achieved by cliff falls and landsliding and is usually stimulated by wave attack and groundwater levels in the coastal slopes. The principle recession mechanisms are falls, slides and flows (Figure 10.2), and encompassing the often dramatic effects of **seepage erosion**. **Falls** occur when material becomes detached from a cliff face; they occur whenever the coast is retreating, but seldom leave a lasting trace. **Slides** involve movement along a basal shear surface and can take a variety of forms : rotational, translational or compound.

Not all clifflines, however, are undergoing rapid rates of retreat. Those formed of the tougher rocks underlying some areas of Britain (Figure 2.6), are retreating very slowly so that landsliding is of limited frequency and often relatively small scale. By contrast, where failures are more frequent and sometimes of a larger scale, the rates of cliff recession are higher; many of the most dramatic examples occur on cliffs developed in soft sedimentary rocks and of weak glacial deposits that form the margins of southern and eastern England.

The coastline in Britain is very long and well known for its varied character, rapidly changing rock type and intensity of marine erosion, so it should come as no surprise that the major areas of coastal recession are correspondingly diverse. Four broad categories of coastal recession can be recognised on the basis of ground-forming materials and type of failure (Figure 10.3; Hutchinson, 1984, Jones and Lee, 1994):

(i) **Major coastal landslides in weak superficial deposits**; The east coast of England from North Yorkshire to Essex is largely developed in thick sequences of glacial till interbedded with sands and gravels, occasionally overlying hard rock cliffs. These deposits can be rapidly eroded by the sea, so that coastal landslides are ubiquitous except where the cliffs are protected by sea defences. For example, the entire 60km length of the undefended Holderness coastline (Humberside) has retreated at average rates of 1.8m pa since 1852. On the North Yorkshire coast the till cliffs can be prone to major dramatic landslide events; the Holbeck Hall failure of June 1993 was the latest example.

(ii) **Major coastal landslides developed in stiff clay**; Stiff clays are particularly prone to landsliding with classic examples occurring along the southern shore of the Thames estuary in Kent. Here cliffs up to 40m high developed in London Clay have repeatedly failed in response to marine erosion, which results in average retreat rates of up to 2m per year.

Although minor failures occur elsewhere along the coast where clay strata outcrop, the largest failures are associated with interbedded clays and sands. There are conspicuous failures of this type on the north coast of the Isle of Wight, especially at Bouldner where the cliffs are up to 50m high. On the coast of Christchurch Bay, at Barton-on-Sea, landslides extend for 5km along cliffs up to 30m high developed in Barton Clay and Barton Sand overlain by Plateau Gravel.

Figure 10.1 A 3-stage cliff recession model.

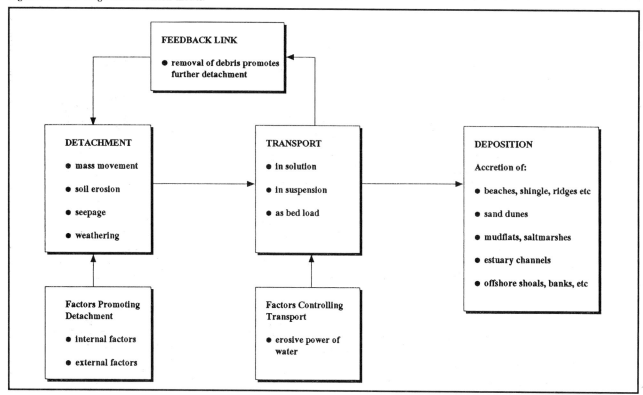

(iii) **Landslides developed in stiff clay with a hard cap-rock**; The largest coastal landslides occur in situations where a thick consolidated clay (shale) stratum is overlain by a rigid cap-rock of sandstone or limestone, or sandwiched between two such layers. Amongst the most dramatic examples is Folkestone Warren, Kent, where the high Chalk Cliffs have failed on the underlying Gault Clay, the Isle of Wight Undercliff and at Black Ven on the west Dorset coast.

(iv) **Landslides developed in hard rock**; Coastal cliffs developed in rocks are continually suffering minor collapse due to basal undermining by the sea. These events are most frequent in the rock clifflines of south-eastern England such as the famous Seven Sisters and Beachy Head Chalk cliffs of East Sussex, which are currently retreating at an average rate of 0.97m a year. Large falls occur on a number of coasts including the Seven Sisters coastline of Sussex, the Chalk cliffs to the north and south of Dover, Joss Bay in Kent, the Triassic sandstone cliffs of Sidmouth, Devon and the Liassic limestone cliffs of South Glamorgan.

Coastal recession leads to land loss. Although individual failures often tend to cause only small amounts of cliff retreat, the cumulative effects can be dramatic. For example, the Holderness coast has retreated by around 2km over the last 1000 years, including at least 26 villages listed in the Domesday survey of 1086; 75Mm^3 of land has been lost in 100 years, (Valentin, 1954).

Cliff recession can be an intermittent process, with periods of little or no erosion separated by rapid and occasionally dramatic landslides, which may remove large sections of cliff in a single event. Major failures are a feature of the till cliffs of north east England. In 1682, for example, the nearby village of Runswick was destroyed by a sudden cliff failure. At Scarborough, large slides have occurred in 1737 and in 1892. The most recent example occurred, on the morning of 4 June 1993 when guests at the Holbeck Hall Hotel woke to discover that a major landslide had occurred on the 70m high coastal cliffs in front of the hotel. Over 60m of cliff was lost overnight, leaving the hotel in a very dangerous position as cracks began to develop in the building. The guests were evacuated during breakfast and the area closed off by the police and the local council. A further 35m of cliff collapsed over the next three days undermining the hotel which gradually toppled over the cliff edge (Clark and Guest, 1994).

Figure 10.2 Landslide types (after Geomorphological Services Ltd, 1987).

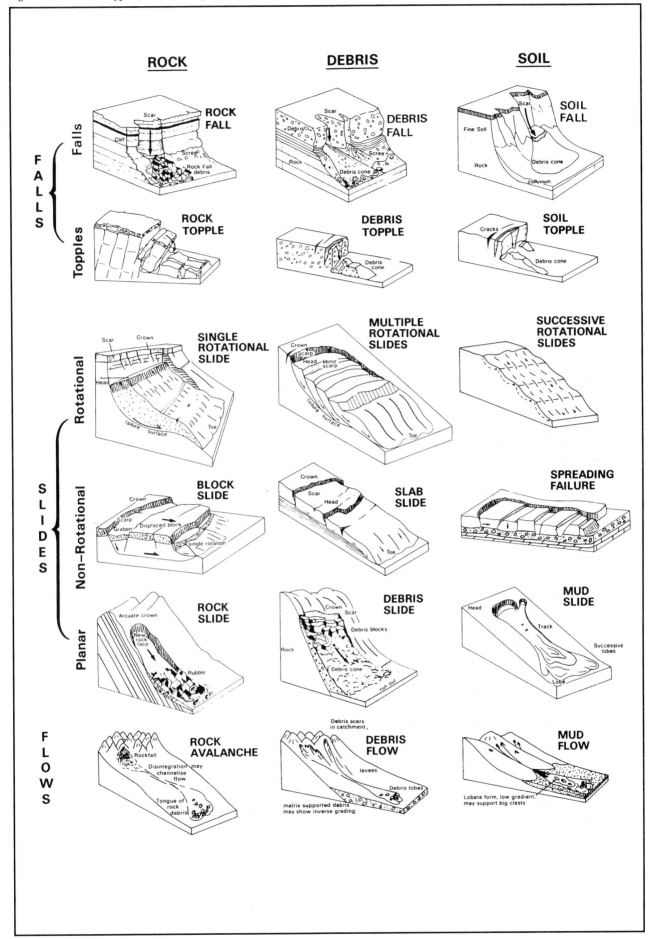

Figure 10.3 Characteristic eroding cliff types.

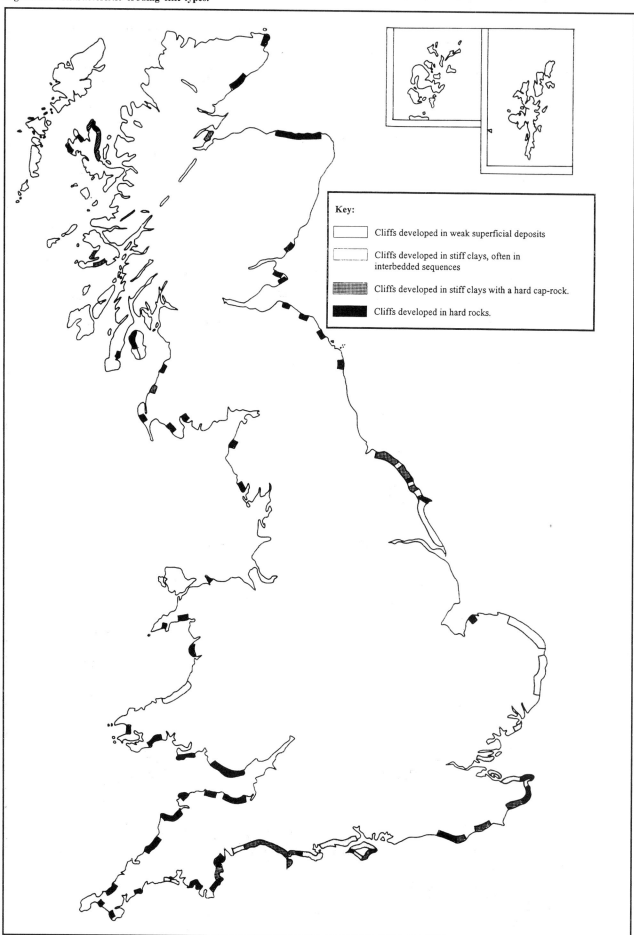

Key:

☐ Cliffs developed in weak superficial deposits

☐ Cliffs developed in stiff clays, often in interbedded sequences

▦ Cliffs developed in stiff clays with a hard cap-rock.

■ Cliffs developed in hard rocks.

Figure 10.4 Geotechnical classification of eroding coastal cliffs.

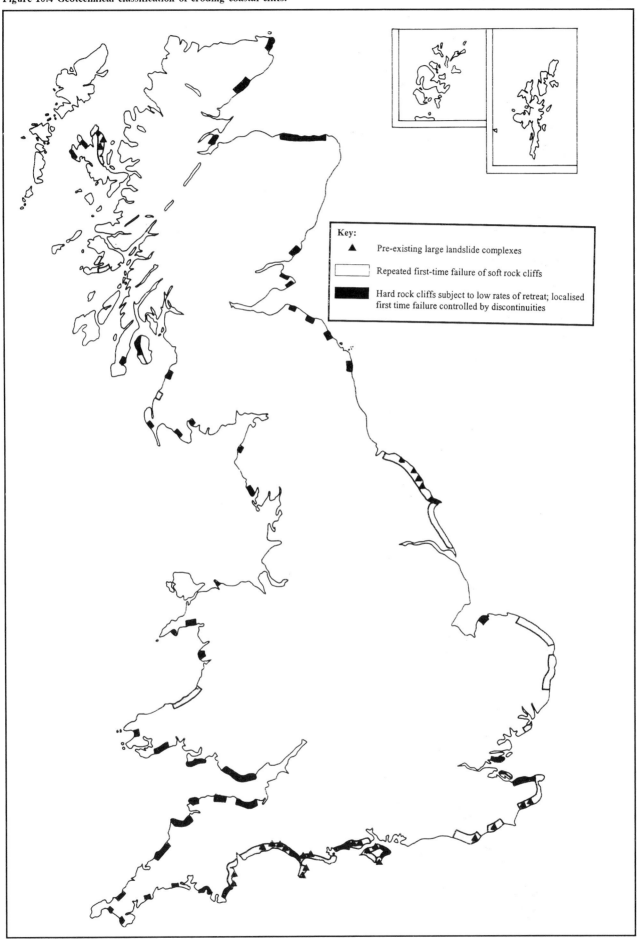

Key:

▲ Pre-existing large landslide complexes

☐ Repeated first-time failure of soft rock cliffs

■ Hard rock cliffs subject to low rates of retreat; localised first time failure controlled by discontinuities

129

The nature of the problem along the south coast is rather different to that in North Yorkshire and Humberside. The generally slower rates of erosion on the sedimentary rocks (as compared with glacial deposits) often means longer intervals between failures and therefore greater sense of security. In addition, mass movements will tend to be slower, creating damage and disruption rather than wholesale destruction. Under these circumstances, quite long–lived communities can come to be threatened, which tends to heighten the sense of local concern and outrage. Nevertheless, locally significant rates of erosion have also been recorded along the south coast and coastal landsliding has been widely reported as affecting many coastal communities including Sandgate, Ventnor, Barton-on-Sea, Swanage, Charmouth, Lyme Regis, Torbay and Downderry in Cornwall.

The small village of Fairlight, East Sussex, is threatened by the retreat of 30m high cliffs which have endangered part of the community. Sections of the cliff have been breaking off at up to 3.5m per annum in the 1980s, threatening 11 houses with destruction by the year 2000. It has been feared that 46 properties could go within a century. In view of the seriousness of the problems facing the community, the Fairlight Coastal Preservation Association was set up to protect the interests of home owners, and lobbied for a coast protection scheme to reduce the rate the erosion. This has now been undertaken at a cost of £2.5 million, involving the construction of a 500m long, 30m wide, concrete armoured berm in front of the cliffs.

Without a doubt, the most extensive coastal landslide problem in Great Britain is at Ventnor, also on the Isle of Wight, where the whole town has been built on an ancient landslide complex (Lee and Moore, 1991; Lee et al. 1991). Although present day coastal retreat is minimal, long–term erosion has helped shape a belt of unstable land which extends almost 1km inland. Contemporary movements within the town have been slight; however, because movement occurs in an urban area with a permanent population of over 6000, the cumulative damage to roads, buildings and services has been substantial. Over the last 100 years about 50 houses and hotels have had to be demolished because of ground movement. In addition, from a consideration of the accumulative damage, including an assessment for repeated road repairs by the Isle of Wight County Council, the total losses in the Ventnor area during the 20 years prior to 1980 have been estimated as exceeding £1.5 million.

Coastal landslides can also lead to loss of life or serious injury in sudden rockfalls or cliff collapses. For example, the popular tourist centres of the Dorset coast have had a series of recent tragedies. On Sunday February 21, 1977, a school party from Warlington in Surrey were studying the geology of Lulworth Cove on the Dorset coast, when they were buried beneath a sudden rock slide. Despite rescue attempts by local ambulance men, the schoolteacher and a pupil had been killed and two more pupils seriously injured, one of whom died later in hospital. This had not been an isolated incident. In 1925, eight workmen, busy extending a road from Bascombe to Southbourne, were buried beneath 100 tonnes of rubble from a rockfall, three of whom died. On August 28, 1971, a nine year old girl was hit on the head by falling rock whilst walking on the beach at Kimmeridge, and later died of her injuries. At Swanage, a schoolboy on a field course was seriously injured by a rock fall in February 1975, and a year later (April 1976) a young boy was killed after being hit on the head by falling rock. In July 1979 a woman, sunbathing on the beach near Durdle Door, was killed when a 3m overhang collapsed. These incidents, and others, led the Chief Inspector of Wareham police to coin the phase "killer cliffs", highlighting the serious danger that rock falls and landslides posed to tourists and educational parties.

The risk associated with cliff recession events is determined by the type of movements which can be expected to occur and their potential consequences. Although a wide variety of factors (e.g. material characteristics, geological structure, pore water pressures, slope angle, etc.) and causes (e.g. coastal erosion, weathering, seepage erosion, high groundwater levels etc.) are important in determining the occurrence of mass movement, it is the effective **shear strength** operating at various depths within a slope and along any pre–existing shear surfaces which is critical in controlling character of failure. Three main groups of eroding cliff can be recognised (Figure 10.4)

(i)   **cliffs prone to first-time failures** of previously unsheared ground, often involving the mobilization of the **peak strength** of the material. Such landslides are often characterised by large, rapid displacements, particularly if there are large differences between the peak and residual strength values. The Holbeck Hall landslide of June 1993 is a good example of this style of failure, with the dramatic movements occurring on an intact coastal slope developed in glacial till.

Repeated failures of unsheared slopes is a common feature around the coast. Erosion of the soft glacial till cliffs of the Holderness coast, for example, involves relatively small first-time failures at a given point every 1–5 years;

(ii) **cliffs prone to failure along pre-existing lines of weakness** such as faults, joints or bedding planes. On many hard rock coasts the potential for failure and cliff recession is controlled more by the presence and pattern of these discontinuities than the strength of the rock mass itself.

(iii) **cliffs prone to reactivation of pre-existing landslides** where part or all of a previous landslide mass is involved in new movements, along shear surfaces where the materials are at **residual strength** and non-brittle. In many inland situations landslides can remain dormant or relatively inactive for thousands of years. However, in the case of coastal landslides, marine erosion removes material from the lower parts of the slopes, thereby removing passive support and promoting further movement. Such failures are generally slow moving, although more dramatic failures can occur.

The importance of this distinction between first-time and pre-existing slides is that once a slide has occurred it can be made to move under conditions that the slope, prior to failure, could have resisted.

Many cliffs present important sources of littoral sediments for landforms such as beaches, sand dunes and mudflats. The annual input of sediment in to the coastal zone from the eroding Holderness coastline is believed to be around $1Mm^3$ of which $0.7Mm^3$ is fine grained material, some of which is carried into the Humber estuary where it contributes to the channel sedimentation described in Chapter 8. The coarser material ($0.3Mm^3$ a year) is moved southwards to Spurn Head and, possibly, across the Humber to the beaches and dunes of the Lincolnshire coast. Disruption of the supply of sediment from eroding cliffs will invariably lead to the starvation of some coastal landforms and, hence, may lead to increased erosion or flood problems elsewhere.

The Causes of Coastal Recession and Landslides

The ultimate cause of all landsliding is the downward pull of gravity. The **stress** imposed by gravity is resisted by the **strength** of the material. A stable slope is one where the destabilising stresses can be overcome by the mobilisation of resisting forces and, therefore, can be considered to have a **margin of stability**. By contrast, a slope at the point of failure has no margin of stability, for the resisting destabilising forces are approximately equal.

Progressively higher values represent more and more stable situations, with greater margins of stability. The higher the value the greater the ability of the cliff to accommodate change before failure occurs. Slope movements are the result of changes which upset the balance between resistance and destabilisation; this ability to withstand change marks the stability state of a cliff:

- **stable**; where the margin of stability is sufficiently high to withstand all transient forces in the short to medium term (i.e. hundreds of years), excluding excessive alteration by human activity;

- **marginally stable**; where the balance of forces is such that the slope will fail at some future time in response to transient forces attaining a certain level of activity;

- **actively unstable**; where transient forces produce almost continuous movement.

This classification makes it possible to recognise that the work of destabilising influences can be apportioned between two categories on the basis of their role in promoting slope failure. These are:

- **preparatory factors** which work to make the slope increasingly susceptible to failure without actually initiating it;

- **triggering factors** which initiate movement.

Cliff recession can be an episodic process with periods of little or no erosion separated by rapid erosion; occasionally dramatic landslides which may remove large sections of coastline in a single event. The 1993 Holbeck Hall landslide at Scarborough, for example, involved around 95m of cliff retreat over a five day period, with 60m lost overnight. The rate of recession is controlled by sequences of events which can range in frequency from 4–5 years (e.g. the Holderness coast) to over 5,000 years (e.g. the Isle of Wight Undercliff). Recession events should be viewed as inherently uncertain because of the complex response of different cliffs to events such as extreme rainfall or storm surges.

The actual causes of cliff recession and landsliding are extremely complex and varied, although they can be considered in terms of **external** factors which increase the shear stress and **internal** factors which lead to a reduction in the shear strength of the cliffs. A detailed analysis of landslide causes can be found in the DoE – commissioned Review of Landsliding in Great Britain (Geomorphological Services Ltd, 1987; Jones and Lee, 1994), although the principal factors influencing coastal landslide activity are:

●   wave energy
●   slope characteristics
●   water regime

Wave action at the base of a coastal cliff can cause oversteepening of the slope and, hence, an increase in shear stress. This is readily apparent on coasts prone to rock falls, with distinctive wave–cut notches often visible. In such circumstances wave action can trigger the failure, but elsewhere slope steepening merely prepares the cliff for failure. This is most dramatically seen at Stonebarrow, Dorset (Figure 10.4; Brunsden and Jones, 1976) where erosion of landslide debris by the sea maintains mudslide activity, whereas removal of support to the rear cliffs by landslide movement downslope, stimulates rotational failures further inland. Over–steepening of the sea cliffs leads to instability in the rear parts of landslide complex, which in turn, eventually causes increased debris supply to the shore. As this landslide debris has to be removed before the bedrock cliffs can be exposed to basal undermining, the result is cycles or pulses of landsliding activity generated by "**waves or aggression**" transferred inland by each phase of cliff retreat.

The importance of the relationship between the rates of basal erosion and debris supply have long been recognised for the London Clay sea cliffs of north Kent and Essex, where three main modes of failure have been identified by Hutchinson (1973).

(i)   balance between rate of debris supply and debris removal results in a scalloped cliffline fashioned by shallow **mudslide** systems;

(ii)   erosion–dominated sites result in straight, steep cliffs with large, deep–seated rotational landslides (e.g. Warden Point, Isle of Sheppey);

(iii)   deposition–dominated situations result in the gradual creation of stability through 'free degradation' which may be permanent if erosion has ceased, as at Hadleigh Castle, Essex (Hutchinson and Gostelow, 1976).

The effectiveness of wave action is controlled by how the waves transform through shoaling as they approach the cliff. Where no shoaling occurs, as when cliffs plunge into deep water, waves are reflected and little erosion takes place. When waves break offshore, on a low beach or shore platform, only limited erosion occurs. Wave action has most effect on cliffs whose base is fronted by a relatively narrow steep beach, where the cliff base experiences the maximum erosive force of the breaking waves. On many coasts, only very short steep waves are capable of progressing across a beach or shore platform without breaking, emphasising the importance of rare, high magnitude waves. Smaller waves may act to remove landslide material, preventing the formation of a protective accumulation of debris at the base of the cliff and, hence, enabling further recession to occur.

Wave energy is not constant around the coast, reflecting the varying exposure to waves of different fetch. The Western Isles bear the full force of the wave energy arriving at the British coast, where the 50 year return period wave height is 35m (Figure 2.4). However, the wave force is quickly attenuated away from the open ocean, declining to around 10m at Dover. Although the east coast is sheltered from the dominant south west waves, this coast suffers from tidal build up (see Chapter 9) and has a 600km fetch from storms from the north. Combined these can cause massive wave energy arriving at the coast as the greater depths close inshore cause waves to lose less of their energy in bed friction or breaking before they reach the shoreline. The tidal range can also be an important factor; on coasts with large tidal ranges the waves are continuously being shifted up and down the shore profile by the movement of the tide, reducing the period over which the cliff foot is exposed to wave attack (Figure 2.10).

Not all cliffs are equally prone to failure or fail in the same manner; rapid erosion and particular types of landsliding are associated with particular types

Figure 10.5 The cyclic evolution of the Stonebarrow landslide complex (after Brunsden and Jones, 1976).

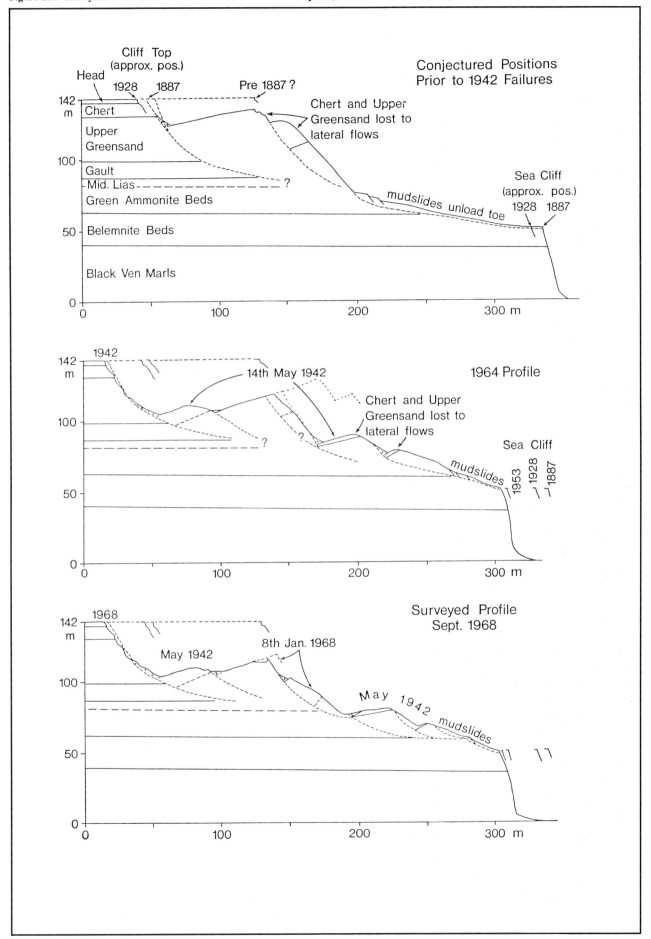

of material. The nature of the cliff materials is, therefore, a fundamental influence on the distribution and nature of cliff erosion problems, especially:

(i) **lithology**; a slope is only as strong as the weakest horizon, often a clay. In general, materials become weaker as the clay content increases and, hence, more prone to failure; this is illustrated by the differences in shear strength between **frictional** materials (e.g. sands and hard rocks) and **cohesive** materials (e.g. clays and mudrocks) shown in Table 10.1.

It is the vertical variations in lithology between one bed of rock and another, or between bedrock and overlying superficial deposits, or even within superficial deposits themselves, that are of crucial importance because they produce the differences in water–bearing characteristics and strength. These factors profoundly influence the siting of instability and largely determine the nature of the resultant landsliding. In alternating sequences of clays and sands or clays/shales and limestones, variations in strength will mean that fairly tough beds of rock will be involved in movements because of the failure of weaker underlying layers. This is well displayed at Stonebarrow, Dorset (Figure 10.4), where strong, rigid beds of Upper Greensand Chert have been involved in rotational landsliding due to failure in the underlying weaker Gault Clay.

(ii) **weathering**; the shear strength of most materials will gradually be reduced over time by the processes of **physical** and **chemical** weathering, (Table 10.2). As the materials are progressively weakened, so the cliff stability can decline towards a potentially unstable condition.

(iii) **hydrogeology**; water content has a major influence on reducing shear resistance because groundwater reduces the amount of contact between soil particles and, hence, the frictional component of shear strength. The frictional resistance to movement depends, therefore, on the difference between the applied **total normal stress** and the pressure exerted by water in the soil pore spaces (**pore water pressure**). This difference, or that part of the normal stress which is effective in generating shear resistance, is known as the **effective stress**.

Pore–water pressures are generally very variable within a cliff, reflecting local variations in lithology. Springlines or seepage zones often indicate the surface outcrop of impermeable materials; these represent areas of concentrated discharge from an **aquifer** and may give rise to zones of high pore–water pressures where the cliff stability is reduced. The combined effects of lithological variation and associated groundwater conditions are a common cause of landsliding, especially along the south coast of England, at sites such as Barton–on–Sea where permeable gravels and sands overlie clay beds.

(iv) **structure**; joints, faults, bedding planes and other **discontinuities** represent lines of weakness within a rock mass and can have a major influence on cliff stability. The dip of the bedding within the rocks is of crucial importance to cliff stability, for the direction of dip defines the amount and direction of water movement in an alternating sequence of sediments and the direction of bedding plane slippage. If the direction of dip is into the cliffline, then the cliffs may be relatively stable. However, if the dip is towards the sea then the cliffs will be potentially unstable because failures can preferentially follow weak strata. The steeper the dips, the more extreme the potential stability or instability.

Changes in the **water regime** as a result of rainfall can quickly affect cliff stability and have been responsible for triggering or reactivating more landslides in Great Britain than any other factor (Jones and Lee, 1994). Examples of landslides caused by high pore–water pressures are widespread throughout Great Britain. The following example is merely intended to illustrate how high pore–water pressures can act as both **preparatory** and **triggering** factors.

Between November 1987 and January 1988 landslide movements occurred at Luccombe Village, Isle of Wight, resulting in severe damage to a number of properties. The village had been built in about 1930 on relatively gentle slopes above a 80m high sea–cliff developed in alternating sandstones and clays of the Lower Greensand. The slopes upon which the village was built are now known to form part of an ancient

Table 10.1 Typical properties of soils and rocks (after Selby, 1982).

Material	Friction Angle Degrees	Cohesion KPa
Loose sand, uniform grain size	28–34	
Dense sand, uniform grain size	32–40	
Loose sand, mixed grain size	34–40	
Dense sand, mixed grain size	38–46	
Gravel, uniform grain size	34–37	
Sand & gravel, mixed grain size	48–45	
Chalk	30–40	
Sandstone	35–45	
Limestone	35–40	
Shale	30–35	
Granite	45–50	
Basalt	40–50	
Very soft clay	12–16	10–30
Soft clay (e.g. lacustrine)	22–27	20–50
Soft glacial clay	27–32	30–70
Stiff glacial clay	30–32	70–150
Glacial till (mixed grain size)	32–35	150–250
Soft sedimentary rock e.g. sandstone, shale, chalk	25–35	1,000–20,000
Hard sedimentary rocks, e.g. limestone, older sandstones	30–40	10,000–30,000
Metamorphic rocks e.g. gneiss, slate	35–45	20,000–40,000
Hard igneous rocks e.g. granite, basalt	35–45	35,000–55,000

Note: Shear strength is a function of cohesion between particles and the frictional resistance, as measured by the angle of friction.

landslide system developed in the Gault Clay and overlying Upper Greensand, and thus the recent movements have involved failure along pre-existing shear surfaces at, or close to, residual strength (Lee and Moore, 1989). Retreat of the sea cliffs at Luccombe is estimated at 0.3m per year. This has had the effect of reducing support to the landslide slopes underlying the village as the coastline retreated inland of the original position of the landslide toe. As a result of this long–term process, the stability of the inland slopes gradually deteriorated. In addition, the development of housing in the area has contributed to a reduction in stability through the disruption of the natural drainage and artificial groundwater recharge by leakage from septic tanks and water supply pipes. It is clear that human activity, together with long–term coastal erosion, have acted as **preparatory factors**, and have produced a situation whereby the slopes within the village have become increasingly susceptible to reactivation. However, the movements in 1987–1988 were triggered by a prolonged period of heavy rainfall during which 638mm fell between September 1987 – January 1988 (the 4th longest wet phase since 1947).

The relationships between rainfall arriving at the ground surface, the generation of pore–water pressures within ground composed of variable materials, and the creation of instability, are extremely complex. A slope experiencing oversteepening by coastal erosion may be subjected to similar rainfall events on many occasions without displaying signs of stress until a point is reached when such an event causes failure because of the changes in slope geometry. Similarly, modification to slopes through the creation of foundations, retaining walls etc., may impede groundwater movements, thereby causing saturation so that the same rainfall events that were harmless in the past now create high pore–water pressures that may lead to failure.

Infiltration of water into a slope varies with the saturation of the ground; when the soil is dry and cracked there may be rapid infiltration whereas much of the rainfall may run–off of saturated slope. Moisture status prior to a specific event is known as antecedent moisture condition and is affected by the pattern of previous rainfall episodes seepage from water bodies such as ponds and leakage of water pipes, storm sewers etc. Such antecedent conditions may be extremely variable and can display both short–term fluctuations and long–term trends. As a consequence, the relationships between rainfall events and slope

Table 10.2 Classification of weathering process.

```
Physical Weathering: Processes of Disintegration

a.    Crystalline processes
      Salt weathering (crystal growth, hydration,
      thermal expansion)
      Frost weathering

b.    Temperature/pressure change processes
      Insolation weathering
      Sheeting, unloading

c.    Weathering by wetting and drying
      Moisture swelling
      Alternate wetting and drying
      Water-layer weathering

d.    Organic processes
      Root wedging
      Colloidal plucking
      Lichen activity

Chemical Weathering: Processes of Decomposition

a.    Hydration and hydrolosis

b.    Oxidation and reduction

c.    Solution, carbonation, sulphation

d.    Chelation

e.    Biological chemical changes
      Micro-organism decay, bacteria, lichens
```

instability may vary considerably both in terms of the size of the triggering event and the lag-time after a rainstorm for the water to infiltrate and raise groundwater levels.

The Identification of Vulnerable Areas

The Review of Landsliding in Great Britain (Geomorphological Services Ltd, 1987; Jones and Lee, 1994) has identified the distribution of recorded landslides around the British coast. The principal areas of landslide activity are and shown on the accompanying 1:625,000 scale thematic maps as "eroding cliffs". The pattern highlights a number of particularly vulnerable geological settings:

Group A; stiff fissured clays and mudrocks where low shear strengths and a high susceptibility to weathering and softening has led to large numbers of failures on oversteepened slopes. The most common forms of landsliding on these materials include single and successive rotational slides,

debris slides and mudslides. The most landslide prone materials in this category are the London Clay, the Gault Clay and Lias Clays of southern and eastern England;

Group B; well jointed, faulted, cleaved and foliated hard rocks in which the pattern of discontinuities provides potential failure surfaces or weak zones within the rock mass. Rockfalls, topples, sagging failures and rock slides are the dominant modes of failure of these rocks, which include Permian Basal Breccias and Devonian limestones in Torbay, the Carboniferous limestones of Wales and Northern England, and the Chalk along the south coast of England.

Group C; the occurrence of sequences of lithologically variable rock types which create potentially unstable conditions. For example, many areas of known landsliding are associated with the presence of thick horizons of impermeable fissured clays or mudrocks overlain by a massive, but well jointed, permeable caprock of sandstone, limestone or volcanic rocks. Multiple rotational slides and compound failures are the dominant forms of landsliding associated with this setting. Classic examples of areas with landsliding promoted by these unstable combinations of rocks include the Upper Greensand overlying the Gault Clay along the Isle of Wight Undercliff, or the same unit overlying the Lias Clays along the West Dorset coast at landslides such as Black Van and Stonebarrow.

Group D; the occurrence of weak superficial deposits either mantling rock cliffs or forming the whole cliff section. The latter settings are particularly prone to rapid erosion, as along the Holderness and Norfolk–Suffolk coasts of eastern England. A combination of small rotational failures and mudslides are the most frequent types of landsliding, although dramatic large failure can develop in particular circumstances, as occurred at Holbeck Hall, Scarborough, in June 1993.

At a local level reliable measurements and predictions of cliff recession rates are needed by both coastal engineers and planners, most notably for:

●   identification of communities at risk from future erosion events;

●   selecting appropriate erosion control techniques, including "soft engineering" methods;

- defining "set back" lines within which development could be affected by coastal erosion during a specific period.

In the past, cliff recession rates have largely been determined from historical data, especially sequences of topographic maps. However, the process is irregular; there may be little or no erosion for a long period, followed by sudden rapid recession in response to a large storm. The "rate" of recession may, therefore, depend on the timescale chosen. Furthermore, some of the main factors such as wave height are inherently unpredictable. This means that it is difficult to predict the exact recession rate for a particular period in the future. These difficulties have been recognised by MAFF who have recently commissioned Rendel Geotechnics and HR Wallingford to undertake a three-year research programme to:

- develop analytical methods of predicting cliff erosion rates for a wide variety of situations around the coast;

- develop a methodology for taking accurate measurements and recording actual recession rates;

- review and evaluate a range of methods for reducing and controlling erosion.

The Effects of Development

Development can have a significant effect on cliff stability, both at the particular site and on adjacent slopes. One of the most serious effects is often the artificial recharge of the groundwater table and, hence, increased pore-water pressures. At Ventnor, Isle of Wight, for example, it has been shown that uncontrolled discharge of surface water through soakaways and highway drains may have contributed to raising the groundwater table to a level where heavy winter storms could trigger movement (Lee and Moore, 1991). In addition, progressive deterioration and leakage of swimming pools and services such as foul sewers, storm sewers, water mains and service pipes can all contribute to cliff instability problems. The example from Luccombe, Isle of Wight, described earlier clearly illustrates this point with leakage from septic tanks and water supply pipes being factors in causing the 1987–1988 ground movements (Lee and Moore, 1989).

Housing development can also affect slope stability through inappropriate excavations to create level plots for buildings. This is best illustrated by reference to the problems encountered during regrading prior to house building along Marine Parade, Lyme Regis, in February 1962 (Hutchinson 1984b; Lee 1992). Around 20,000m^3 of material was removed from the slope, which was frequently affected by shallow landsliding. A few days after the earth moving operation had finished a large, deep-seated slide developed which moved several metres in a few minutes. It was established subsequently that the failure had occurred on a pre-existing landslide shear surface. The regrading, which was designed to improve stability against shallow landsliding, had in fact removed support from the toe of an unsuspected, potentially unstable, large slide (Figure 10.6).

Excavations can also lead to landsliding on previously unfailed slopes, especially when the materials are very sensitive silts and clays prone to liquefaction. The best example in Britain is the large failure that occurred in June 1975 during construction of the Portavadie Dry Dock in western Scotland (Clarke et al 1979). The Dock was being excavated through a variable sequence of Quaternary deposits and recent marine laminated silts, when a slide occurred which caused slurried laminated silts to move up to 270m away from their original position on the excavated face. The slide is considered to have been caused by a chain reaction from an initial relatively small slip on the steep excavated face. This first failure produced a shock wave which passed through the soil mass causing loss of structure and large-scale rapid movement.

Coastal instability can also be exacerbated by the disruption of sediment transport, as occurred at Folkestone Warren following harbour construction (Hutchinson, 1969; Hutchinson et al 1980). The 130–155m high cliffs of Chalk overlying Gault Clay have experienced increased landslide activity over the last 100 years. This is attributed to decreased beach volume and the accelerated removal of the toes of previously displaced landslide blocks. Twelve major slips have been recorded since 1765, with a particularly dramatic failure in 1915 when virtually the whole of the Warren was involved in seaward movement, including one large block which displaced the Folkestone to Dover railway line by up to 50m. Although coastal landsliding would have been reactivated at this location some 6000–7000 years ago by the post-Glacial rise in sea-level, this most recent phase of pronounced failure was probably a

137

Figure 10.6 The 1962 Marine Parade landslide at Lyme Regis (after Lee, 1992).

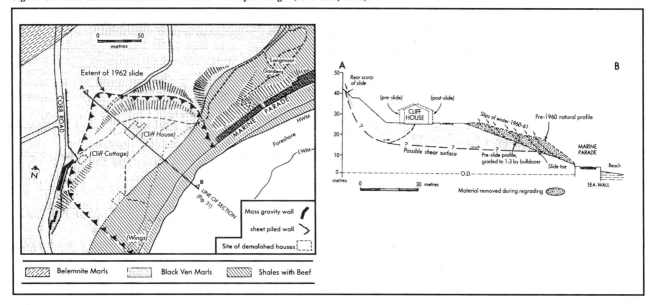

consequence of the expansion of the Folkestone Harbour facilities over the years 1810–1905. In particular, the extension of the main pier resulted in the disruption of wave and current induced littoral drift of sand and shingle eastwards through the area, leading to the build up of trapped material west of the pier and beach shrinkage through undernourishment to the east, at the foot of the Warren. Reduced beach volume would have led to increased wave erosion of the toes of pre–existing rotated blocks, which slid seawards thereby removing support from the base of the cliffs. The decrease in passive support would have created conditions of marginal stability, with actual failure taking place at times of high groundwater levels following rainfall. Today, stability is maintained by the existence of major toe–weighting and erosion control structures.

These examples can only illustrate the variety of ways in which development can affect coastal instability. In many areas, the human influence can be a key factor in understanding landslide problems:

"In Britain instability problems are not Acts of God; unpredictable, entirely natural events .... Man's role in initiating or reactivating many slope problems should not be underestimated" (Jones and Lee, 1995).

The Significance for Conservation

Some landslides and retreating cliffs are of particular importance to earth science research and training, and have been recognised as GCR sites. At present some 9 coastal landslide sites in Great Britain have been identified (Table 10.3), most of which lie within existing SSSIs.

Landsliding can also provide scenic attractiveness. On the coast, cliffs are shaped by and dependent on landslide and recession processes. They are amongst the nation's greatest assets with many safeguarded by the protection afforded by their inclusion in National Parks and AONB's or through their status as heritage coasts. At present around 1525km of coast in England and Wales has heritage coast status, with public enjoyment encouraged by the provision of recreation activities that are consistent with their conservation of their natural scenery and heritage features.

Many hard rock cliffs are renowned for providing prime breeding grounds for seabirds, with cliffs from Flamborough Head north to Dunnet Head, Cape Wrath to Land's End and the Northern and Western Isles containing the bulk of Europe's seabird population. Indeed, over 20% of the world's population of razorbills nest around the Great Britain coast. Many actively eroding coastal landslides give rise to important habitats as colonising species thrive on the environmental change resulting from instability. Numerous threatened species are found in such settings; hoary stock (Matthiola incaria) found only on eroding chalk cliffs, the Scottish primrose (Primula scotica) on cliff tops in Orkney.

Cliffs are also of great value because of the exposures of geological features. This is reflected in the large number of geological SSSIs, especially

Table 10.3 Coastal mass movement sites of conservation value in Great Britain (Source: English Nature).

SITE	GRID REFERENCE	DESCRIPTION
Blacknor Dorset	SY678714	The cliffs on the western coast of the Isle of Portland exhibit probably the best British examples of slab failures. The island is traversed by numerous NE–SW trending joints in the Purbeck and Portland beds, many of them widened by slipping over the underlying Kimmeridge Clay and Portland Sand.
Black Ven Dorset	SY347927	This is a complex and active multiple landslip. The site is important for its demonstration of movement in arenaceous flows of cohesionless remoulded material due to copious water supply from the Cretaceous strata. The addition of more water supplies as flows progress downslope ensure that they reach the beach at the foot of the cliff.
Axmouth – Lyme Regis Devon	SY234899	This is the most renowned area of landslipping in Great Britain. The very large Bindon Landslip of 1839, in the centre of the area, brought the site to public prominence. The features of the Bindon slip are now much obscured by vegetation, but the major topographic elements are clearly discernable from the air.
Folkestone Warren Kent	TR255384	This area of coastal landslides has been more intensively studied than any other of comparable size in Great Britain, because the main Folkestone–Dover railway line was displaced by slipping in 1915. The site has suffered twelve major slips since 1765, and is now protected by a complex of coastal defence works.
Trimingham Norfolk	TG278390	The Norfolk coast shows two areas of particularly impressive rotational slumping affecting Pleistocene deposits. The Trimingham coast is the finest site of slumping of weak, unconsolidated sediments in Britain. Huge collapses of the cliffs continue to occur, in places breaking through an elaborate set of coastal defence works.
Higland (Raasay) Highland	NG587391	This very large, highly unstable area of landslips, involves a thick series of Lower and Middle Jurassic rocks on the eastern side of the Isle of Raasay. The east-facing cliff at the rear of the landslips is about 70m high; at its foot lies a hollow some 500m in breadth after which the land rises as a broad ridge transected by a labyrinth of open fissures of unknown depth. The toe of the landslip is being actively eroded by the sea, and movements have been recorded during the present century.
Quirang (Skye) Highland	NG443665	The Quirang Landslide is the largest in Great Britain, extending 2,130m from the back-face to the coast. It is formed in Tertiary lavas and underlying Jurassic sediments. In front of the back-face is a cluster of high lava pinnacles, and further downslope are numerous slipped lava masses forming hills up to 120m high.
The Storr (Skye) Highland	NG504537	This post-glacial slide in Tertiary lavas is 1,500m long, with a back-face over 600m high and its development is seen to be related to the dolerite sills, which cut the underlying Jurassic strata. Four distinct segments of sediments and overlying lava have successively moved seawards; those furthest upslope from ridges bearing lines of spectacular lava pinnacles up to 49m high, the outstanding examples of this kind in Britain.
Warden Point Kent	TQ961736	At Warden Point, a series of particularly impressive, deep-seated, rotational landslips, bench-shaped in plan, occur in London Clay. Characteristically each slip extends along the coast for distances between four and eight times the cliff height. The back-tilted blocks produced by failure are broken down by shallow slides and mudflows, the debris being removed by marine erosion. This in turn results in progressive steepening of the cliff, and thus further landslipping. The best locality in Britain to observe this cycle of landslip and coastal erosion typical of soft coasts.

on the eroding soft rock cliffs of England from Flamborough Head to the Exe estuary. Such sites include international reference localities for vast periods of geological time, such as the Bartonian Stratotype between Highcliffe and Milford Cliff in Hampshire, and provide opportunities for teaching and research. The oil industry, for example, uses these exposures as training grounds for their geologists who have to recognise similar underground oil field structures from sparse borehole and geophysical data.

Summary: The Significance for Planning and Development

Erosion of coastal cliffs can present significance hazards to development, especially on soft rock coasts where the rates of recession are high. Here, the cumulative loss of land, cliff top properties, services and infrastructure are problems that are experienced on many coastlines. However, the nature of the problems that are encountered vary according to the geological setting and the type of rock fall or landslide event which produces cliff recession. Problems can range from the effects of slow ground movement on unstable coastal slopes, as in Ventnor, Isle of Wight, to the threat to public safety from cliff falls on hard rock coasts. Cliff recession, however, can be an intermittent process with periods of little or not erosion separated by rapid and occasionally dramatic landslides which may remove large sections of coastline in a single event. The 1993 Holbeck Hall landslide at Scarborough, for example, involved around 95m, of cliff retreat over a five day period, with 60m lost overnight.

The causes of cliff recession are extremely complex and varied, although most events are promoted by a combination of wave attack and groundwater levels within the cliff. The effect of development and human activity should, however, never to underestimated, with many cliff failures arising as a result of:

- uncontrolled discharge of surface water through soakaways and highway drains;

- progressive deterioration and leakage of swimming pools and services such as foul sewers, storm sewers, water mains and service pipes;

- inappropriate excavations to create level plots for building, especially at the foot of a cliff;

- disruption of sediment transport by groynes and breakwaters, leading to starvation of beaches in front of cliffs and accelerated erosion.

The last point highlights the important role that cliff recession can have in supplying littoral sediment to beaches, sand dunes and mudflats on adjacent stretches of coastline. Disruption of this sediment supply, by the use of erosion control techniques, may lead to an increase in erosion or flood risk elsewhere, as described in Chapters 8 and 9.

Coastal cliffs are frequently of great conservation value as breeding grounds for seabirds, or by supporting important habitats. Cliffs are also of great value because of the exposures of geological features such as important fossil beds or marker horizons for rocks of a particular age. Continued erosion is necessary to ensure that these features remain accessible for teaching and research purposes. Some coastal landslides are also of particular importance as teaching and research sites for mass movement studies and, consequently, have been designated as SSSIs.

Coastal landsliding and cliff recession are important considerations for both planners and developers. Indeed, development in unsuitable locations can lead to a range of problems from adverse effects on the stability of adjacent land to calls for publicly funded protection measures and the consequent effects on conservation or coastal defence interests elsewhere. The planning system clearly has an important role in minimising the risks associated with coastal instability through:

- guiding development away from unsuitable locations. This may involve establishing "set-back" lines within which development could be affected by erosion over a specified period;

- ensuring that development does not initiate or exacerbate instability problems on adjacent land, by specifying appropriate site drainage requirements and limiting slope excavation during development etc.;

- ensuring that the precautions that are taken to minimise risks from cliff instability do not lead to starvation of sediment supply to other important coastal sites and, thereby, increase the level of risk elsewhere;

- ensuring that development does not have adverse effects on local amenities and conservation interests.

In England and Wales, planning advice on coastal landslide and recession issues is contained within

PPG14 Development on Unstable Lane (DoE 1990) and PPG20 Coastal Planning (DoE, 1992). No comparable advice has been issued by the Scottish Office.

Chapter 10: References

Bray M.J. 1992. Coastal sediment supply and transport. In R.J. Allison (ed.) Coastal Landforms of West Dorset. Geologists' Association Guide, 47, 94–105.

Brunsden D. and Jones D.K.C. 1976. The evolution of landslide slopes in Dorset. Philosophical Transactions of the Royal Society A, 283, 605–631.

Clark A.R. and Guest S. 1994. Holbeck Hall landslide: coast protection and cliff stabilisation. Proc. MAFF Conference of River and Coastal Engineers, Loughborough.

Clark A.R., Hawkins A.B. and Gush W.J. 1979. The Portavadie Dry Dock, West Scotland: a case history of the geotechnical aspects of its construction. Quarterly Journal of Engineering Geology 12, 301–317.

Department of the Environment 1990. PPG 14. Development on Unstable Land. HMSO.

Department of the Environment 1992. PPG 20. Coastal Planning. HMSO.

Geomorphological Services Ltd. 1987. Review of Landsliding in Great Britain. Reports to DoE.

Hutchinson J.N. 1969. A reconsideration of the coastal landslides at Folkestone Warren, Kent. Geotechnique, 19, 6–38.

Hutchinson J.N. 1973. The response of London Clay cliffs to differing rates of toe erosion. Geologia Aapplicata a Idrogeologia, 8, 221–239.

Hutchinson J.N. 1984a. Landslides in Britain and their countermeasures. Journal of the Japan Landslide Society, 21, 1–24.

Hutchinson J.N. 1984b. An influence line approach to the stabilisation of slopes by cuts and fills. Canadian geotechnical Journal, 21, 363–370.

Hutchinson J.N. and Gostelow T.P. 1976. The development of an abandoned cliff in London Clay at Hadleigh, Essex. Philosophical Transactions of the Royal Society A, 382, 557–604.

Hutchinson J.N., Bromhead E.N. and Lupini J.F. 1980. Additional observations on the Folkestone Warren landslides. Quarterly Journal of Engineering Geology 13, 1–31.

Jones D.K.C. and Lee E.M. 1994. Landsliding in Great Britain. HMSO.

Lee E.M. and Moore R. 1989. Report on the study of landsliding in and around Luccombe Village. HMSO.

Lee E.M. and Moore R. 1991. Coastal landslip potential assessment, Ventnor, Isle of Wight. Report on DoE.

Lee E.M. 1992. Urban Landslides. In R.J. Allison (ed). The Coastal landforms of the Dorset Coast. Geologist's Association Guide No. 47.

Selby M.J. 1982. Hillslope materials and processes. Oxford University Press.

Valentin H. 1954. Der landverlust in Holderness, Ostengland von 1852 bis 1952. Die Erde 6, 296–315.

Chapter 10: Suggested Reading

Bromhead E.N. 1986. The stability of slopes. Surrey University Press.

Jones D.K.C. and Lee E.M. 1994. Landsliding in Great Britain. HMSO.

11 Coastal dunes: Wind blown sand

The Nature of the Problems

Coastal dunes are accumulations of blown sand, formed mainly during the period of sea level rise that followed the end of the last glaciation. The Sefton dunes, for example, are believed to have developed in a main phase of dune building between 4600–4000 years ago (Plater et al, 1991; Innes and Tooley, 1991), derived from foreshore and nearshore sand deposits. Since this time there have been marked phases of dune instability, characterised by migrating sands, including around 3000 years ago and during the 18th century.

Five main types of coastal dunes can be recognised around the British coast:

(i)   Offshore island dunes; developed on barrier islands as linear features reflecting the direction of longshore drift. Examples include Blakeney, Norfolk and Morrich More (Ross and Cromarty);

(ii)   Prograding dunes; formed on an open coast where there is an abundant supply of sand either from longshore drift in two directions or from a very shallow sandy shore. Examples include Winterton Ness, Norfolk and Barry Links (Angus);

(iii)   Spit dunes; formed on sandy promontories at the mouths of estuaries. Examples include Studland dunes, Dorset and the Sands of Forvie, Grampian;

(iv)   Bay dunes; developed in bays within the shelter of rock headlands, forming a half-moon shape in the beach and outer dune zone. They are characteristic of the rocky, indented coasts of south west England, Pembrokeshire and Scotland;

(v)   Hindshore dunes; formed on extensive sandy coasts where the prevailing wind is onshore, driving sand inland for considerable distances as a series of dune ridges or mobile parabolic dunes. Examples include Braunton Burrows in Devon, Newborough Warren in Anglesey and Culbin Sands in Grampian. These dunes are also known as the **machair** characteristic of the western and northern coasts of Scotland.

Migration of sand dunes is not a significant problem in Great Britain, but it once was. There is considerable historical evidence to suggest that wind blown sand was a major hazard, especially during the 15th to 17th centuries. Table 11.1 lists some of the best known examples. The range of impacts have included:

●   burial of farms and buildings;
●   rivers blocked and forced to change course;
●   ports left inland.

Perhaps the most memorable series of events was the Culbin Sands disaster of 1694 and following years. At that time, 16 fertile farms covering some 20–30km^2 near Findhorn and Forres on the Moray Firth were overwhelmed in a single violent storm. The whole area including the mansion house was buried by up to 30m of loose sand. From 1694 to 1704 there were frequent periods of blowing sand and the area remained a desert of shifting sand for 230 years until it was successfully afforested by the Forestry Commission during the 1920s. Much of the damage is believed to have been the result of a single violent storm, probably in late September or October, 1694:

"At first only fields were invaded (by the sand). A ploughman had to leave his plough, while reapers left their stooks of barley. When they returned, both plough and barley were buried for ever. The drift then advanced upon the village, engulfing

Table 11.1 Examples of major sand migration events in Great Britain.

Date	Site	Comment
1316	Kenfig, near Port Talbot	Storms causing sand dune movement closed the medieval port of Kenfig. Further events between 1344-1480 finally buried the former Roman coast road.
1385	Harlech	Around this date sand dunes formed, enclosing and protecting the flat area known as Morfa Harlech, closing the Medieval port of Harlech.
1401-1413	Forvie, Grampian	Medieval town buried by sand (now 30m high dunes). Forvie dune advanced 50-250m to the north during the 1413 storm and advanced a further 200m before the end of the 15th century. The storm corresponded with extreme low tides.
1600	Rattray, Grampian	From 1600-1720 the inlet of Strathbeg was buried by sand; it is now the Loch of Strathbeg. Rattray harbour was choked with sand.
1663	Nairn, Grampian	Moving sands threatened to cover the town.
1676	Culbin Sands, Grampian	Culbin estate covered by blown sands to a depth of 0.7m. The source was coastal dunes where marram grass had been eaten by sheep and cattle.
1694	Culbin Sands, Grampian	In the autumn of 1694 16 fertile farms covering 20-30km^2 were overwhelmed in a violent storm. The whole area was buried by up to 30m of sand.
1697	North Uist	Archaeological site buried by drifting sand, carried from shore.
22.10.1702	Findhorn, Moray Firth	Severe drift of loose sand in the Culbin area blocked the R. Findhorn and forced it to change its course.
1739	Sefton	Village of Ravenmeols buried by a great sandstorm.
1794	Happisburgh, Norfolk	Church buried by sand in a storm.
26.12.1862	Happisburgh, Norfolk	The remains of the ancient village of Eccles, which had been buried in sand in the 17th and 18th centuries, was exposed by the exceptionally strong winds.
1870s	Alnmouth	Between 1866-1897 a belt of sand dunes developed on the south side of the estuary.

cottages and the laird's mansion. The storm continued through the night, and next morning some of the cottars had to break through the backs of their houses to get out. On the second day of the storm, the people freed their cattle and fled with their belongings to safer ground. Their flight (southeastwards or eastwards) was obstructed by the river Findhorn: since its mouth had been blocked by the drifting sand, its waters rose until it could force a new passage to the sea." (original account quoted from Edlin, 1976).

The village of Culbin was buried by the storm, although over the following centuries buildings have been released from the sand only to disappear again. The loss of land on the fertile Moray plain would probably have been valued at £15M at current prices, with buried property, lost crops probably raising the overall total to around £25M (Lamb, 1991). In addition, the loss of natural coast protection that had been provided by the dunes before they migrated inland led to the destruction

of the nearby town of Findhorn during a storm around 1702.

It is believed that the erosion on the Moray plain was the result of winds of 100-130 knots (50-65m/sec) blowing inland from the NNW, accompanied by spring tides which probably caused the sea to erode the dune faces, exposing the dry sand cores to wind scour. Another factor may have been the destabilisation of the dunes by harvesting of marram grass for local industries. Indeed, an Act of the Scottish parliament in 1696 legislated against the pulling of marram grass for thatching (Lamb, 1991).

Blown sand problems do still occur, but tend to be localised and considerably less dramatic. The best known examples include Braunton Burrows and Sefton in England, and the mobile sand sheets at Balmedie, Foneran and Forvie in Scotland. At Braunton, for example, the central dunes migrated 122m inland between 1885 and 1957, although they have now been largely stabilised by vegetation. In the early 1900's mobile dunes threatened the Liverpool–Southport railway at Sefton (Jones et al, 1991).

As was stressed in Chapter 9, sand dunes have an important role in providing "natural" defence against tidal flooding, by forming a barrier of high ground in front of a low lying coastal plain. In addition, during severe storms marine erosion of the seaward dunes can mobilise significant quantities of sediment, much of which may accumulate on the foreshore where it helps dissipate wave energy and, hence, reduce the potential for further erosion. Following the storm, much of the "damage" to the dunes will be repaired naturally, as sand is blown back off the foreshore. This ability can be significantly reduced as a consequence of sand and gravel extraction from dunes or the foreshore, leading to an increase in flood risk to low lying land.

The Causes of Dune Migration

Wind blown sand problems are generally associated with three main factors:

- **erodibility** of the sand
- **erosivity** of the wind
- **land management**

Dry, loose sand is considerably more vulnerable to wind erosion than damp sand. Problems tend to occur, therefore, when the seaward faces of dunes are eroded by wave action, exposing the core of the dune. Extremely low tides or periods of relatively low sea level may also expose large areas of loose sand which, if dried by strong winds, can be a source of blown sand. The wind speed and direction are critical for determining the erosivity of a particular storm. Whilst sand can be moved by relatively light winds, the significant events outlined in Table 11.1 are probably associated with severe storms with wind speeds of up to 100 knots or more (over force 12 on Beaufort scale).

Land management is a major factor in both preventing and initiating wind–blown sand problems. Indeed, afforestation of the Culbin Sands, Morrich More and other mobile sand sheets has proved successful in stabilising the dune systems. Elsewhere the planting of marram grass has reduced the local problems associated with wind blow that occur in many dune systems. The importance of these stabilisation measures is

reflected by the widely recognised dangers of uncontrolled marram–cutting; legislation was passed in Scotland after the Culbin Sands disaster and in 1742 an Act of Parliament was laid down "for the more effectual preventing of the cutting of Star or Bent". Indeed, the cutting of star grass (marram) was a serious offence in the 19th century, punishable by fines, whipping or hard labour.

Identification of Vulnerable Areas

Information about the nature and extent of sand dunes around Great Britain has been compiled by JNCC as part of the Sand Dune Vegetation Survey (e.g. Dargie, 1993; Dargie, in press; Radley, in press). Although these areas are sites of potential wind blown sand hazards it is wrong to overstate the magnitude of the hazard and the level of associated risks. As described earlier current problems tend to be localised and relatively minor, largely influenced by land management practices. It is interesting to note, however, that storms of the intensity described for the Culbin Sands disaster do still occur (the 16 October 1987 hurricane produced gusts of over 100 knots), but do not appear to initiate significant sand movement.

To an extent the general location of the dune systems will influence their susceptibility to wind erosion (Doody, 1989):

(i) **exposed west coast** dunes; on the **open coast** prevailing and dominant winds are onshore and largely from the same direction. This combination leads to large expanses of dune that can extend inland. The seaward transition is marked by a narrow steep–faced dune, behind which the progressive landward movement of sand produces a flat sand plain or machair.

(ii) **sheltered west coast** dunes; typically undulating forms develop. Local conditions favour longshore movement and may lead to the rapid formation of spits. Examples include Morfa Harlech and Whiteford Burrows in Wales.

(iii) **exposed east coast** dunes; the prevailing winds are offshore and dominant winds onshore. Dunes are more likely to show evidence of seaward migration. At Morrich More, for example, the movement is offshore. However, severe storms can produce winds directly onshore and may result in major movements, as occurred at Culbin Sands in 1694.

The Effects of Development

Although there may be temporary sand blown problems during construction, built development generally will tend to stabilise dune systems. For example, on the Sefton coast, the mobile dunes that threatened the Southport–Liverpool railway around 1900 were levelled in 1923 for housing (Gresswell, 1953); no significant problems have followed. However, the removal of dune sand for use in construction can cause instability in sensitive dune systems. This may not necessarily lead to major wind blow events, but can lead to a degradation of natural coastal defences and, hence, an increase in flood risk to the inland areas (see Chapter 9).

Golf courses represent one of the major uses of sand dunes in Britain. Over a third of all dunes have been partly modified for this activity (Ranwell, 1975). The positive management of the dune areas, with the encouragement of fairway and "rough" grasses has tended to reduce wind blow problems and provide suitable conditions for the survival of dune vegetation. Recreational use can also lead to destabilisation of dunes. Indeed, concern about the impact of recreation led the Countryside Commission for Scotland to commission a survey of all the beach and dune systems in Scotland (Ritchie and Mather, 1984; Ritchie, 1985). Particular problems can be caused by:

- access to beaches from car parks
- holiday accommodation
- trampling of dunes
- off road vehicles

On exposed west coast sites the effects can be severe with major "blow-outs" and loss of surface vegetation and sand.

Dunes and beaches may contain valuable sand and gravel resources. However, their removal for the construction industry can adversely affect the rate of coastal erosion, and the vulnerability to flooding of adjacent land. In Scotland, large scale sand extraction is common on the dune and beach complexes of the Orkney and Shetland Islands (e.g., the Bay of Quendale) and, to a lesser extent, around Brodick and Girvan on the Clyde coast. It has been estimated that sand extraction has had an adverse effect on 16% of all beach complexes in Scotland (Ritchie and Mather, 1984).

The Significance for Conservation

Sand dunes are of major conservation value. Their wildlife value is considerable with dunes supporting, for example, over 90% of the breeding population of natterjack toad and a rich variety of plant species. Over 120 different characteristic assemblages of vegetation are recognised in dunes (Table 11.2; Radley, 1993) making it one of the most varied habitats in the country; five nationally rare or scarce plants and animals only live in dunes, including dune gentian (Gentiana ulignosa) and hilliborine (Epipactis liptochila).

Dunes are also of considerable value for their geomorphological interest. In Wales, for example, dunes form a major component of seven out of the twelve coastal geomorphology sites selected by the Geological Conservation Review (Table 11.3) including:

(i)   Carmarthen Bay; the Pendine–Laugharne dunes are a classic site for studying the eastwards extension of a dune system, protecting the abandoned rocky cliffs.

(ii)   Morfa Harlech; several sub–parallel dune ridges have developed since the 14th century, leaving the former port of Harlech stranded inland. It is significant for the relationship of the ridges to wave energy inputs from local rivers and the seabed.

(iii)   Morfa Dyffryn; the dunes contain large slacks and fine examples of dune migration. It is a good site for demonstrating dune–beach interactions.

(iv)   Ynyslas; the dunes and sand spit are important for geomorphological studies of estuarine sedimentation.

Sand dunes rely on the repeated cycles of wind erosion and stability, together with marine erosion of the seaward dunes, to maintain a full sequence of successional stages of habitats. In this context, change is a natural process with the stability imposed by many dune management techniques acting to preserve rare species rather than encouraging successional change. It is now recognised that some wind erosion is essential for maintaining dune landscapes and can actually be to

Table 11.2 Major sand dune (SD) vegetation communities (after Radley, 1993; Dargie, 1993).

Standline communities; The SD2 (Honkenya peploides/Cakile maritima) community has a nation-side distribution but with major areal development in Norfolk, Cumbria and the Scilly Isles. In contrast, SD3 (Matricaria maritima/Galium aparine) is found only in Merseyside and Northumberland. Both are present in Scotland. SD4 (Elymus factus) occurs on low dunes all around the Scottish coast.

Mobile dune communities; The SD4 (Elymus factus ssp. borealis atlanticus) foredune community is restricted mainly to the east coast, as is SD5 (Leymus arenarius), which is found only as far south as the Wash. Much more widespread is the SD6 (Ammophila arenaria) community, which dominance most mobile dune belts.

Semi-fixed dune communities; dominated by A. arenaria sub-communities. SD7a (A. arenaria/Festuca rubra 'typical' sub-community) is mainly restricted to Northumberland and Norfolk, but with small areas in the south and west. SD7b (Hypnum cupressiform) has a wide range but is absent on the south coast. SD7c (Ononis repens) and SD7d (Tortula ruralis) both occur nationwide, although SD7c is more abundant on the west coast. In contrast, SD7e (Elymus pycanthus) is largely restricted to the east coast.

Fixed dune grassland communities; dominated by Festuca rubra (SD8), Ammophila arenaria/Arrhenatherum (SD9) and Carex arenaria/Festuca ovina/Agrostis capillaria (SD12) communities. SD8 contains six sub-communities: SD8a ('typical') is the most widespread with SD8b (Lazula compestris) being restricted to the north and west. SD8c (Tortula ruralis) is predominantly western with isolated east coast appearances to the north of the Wash. Both SD8d (Bellis perennis/Ranunculus acris) and SD8e (Prunella vulgaris) are restricted to the north and west. SD9a ('typical') is widespread except in the south and southwest, whilst SD9b (Geranium sanguineum) is exclusively found in the northeast. SD12a (Anthoxanthum sub-community) occurs nationwide while SD12b (Holcus lanatus) is restricted to the north and east.

Fixed dune sand sedge dominated communities; dominated by Carex arenaria (SD10) and C. arenaria/Cornicularia aculeata (SD11). The SD10a (Festuca rubra) sub-community is widespread, but SD10b (F. ovina) is largely restricted to Norfolk and Kent. SD11a (A. arenaria sub-community) has a nationwide distribution, but with a high concentration in Norfolk. The SD11b (F. ovina) sub-community is confined to East Anglia.

Dune slack communities; dominated by Salix repens (SD13, SD14, SD15, and SD16) and Potentilla anserina/Carex nigra (SD17). The SD13a sub-community (Poa annua/Hydrocotyle vulgaris) is nationally rare while SD13b (Holcus lanatus/Festuca rubra) is restricted to Northumberland and Merseyside. The four sub-communities of SD14 (S. repens/Campylium stellatum) have 75% of their national extent at Braunton Burrows, while those of SD15 (S. repens/Calliergan cuspidatum) are restricted to the west coast and a small area in Northumberland. They are generally absent in Scotland.

the benefit of dune management; it can save costly stabilisation measures in non-essential areas and help create a diverse landscape. Research in the Netherlands has shown that many blow-outs can become stabilised without human help, because the internal dynamics of the dunes act to curtail the effect of the wind (Jungerius et al 1981; Jungerius, 1989). Continued wind erosion is, of course, essential for maintaining the geomorphological interest of many dune systems, especially those where there are research opportunities to study the sediment transport interactions between the dunes and adjacent beaches. Finally, it is important to highlight the educational value of dune systems; many are major national centres for the teaching of ecological principles and earth science research, such as the Ynyslas Field Centre in Wales.

Current coastal conservation attitudes recognise the value of dunes as dynamic environments and, hence, the benefits that can be associated with continued wind erosion:

"It is tempting to react to an eroding dune system by attempting to stabilise it. This is to lose sight of ..... cyclicity, and the fact that it is the very instability of dune systems that sustains their wildlife and geomorphological functions. Stabilised dunes lose there value, and it is fruitless to attempt to sustain them to preserve their wildlife and geomorphological characteristics." (Stevens, 1992).

Summary: The Significance for Planning and Development

Wind blown sand does not constitute a significant hazard around the British coastline. This has not always been the case, with the combination of extreme wind speeds in excess of 100 knots and poor land management practices, such as harvesting marram grasses, led to severe dune migration events. The Culbin Sands disaster of around 1694 is probably the most famous event of this kind, and resulted in damages which, at today's prices, probably exceeded £25M. It is important to remember that the potential for dune migration exists, fortunately it is not realised as the risk of a repetition of such events has been minimised by

147

Table 11.3 Coastal dune GCR sites in Wales.

SITE	STATUS	COMMENT
Newborough Warren, Gwynedd	SSSI/GCR Site	Comprises a major dune system with active and fixed sections. The dune forms are varied in character and provide an excellent range of features of different ages.
Tywyn Aberffraw, Gwynedd	SSSI/GCR Site	An area of blown sand and dunes occupying a confined valley site, including individual parabolic dunes on a sand plain. There is little supply of sand to the dunes and the site provides an excellent opportunity for the study of beach–dune relationships.
Morfa Dyffryn, Gwynedd	SSSI/GCR Site	Spit developed across the mouth of Afon Artro, but now links morainic hills to the mainland. Shoreline comprises dunes with examples of migration and large slacks. It is an important member of a suite of west coast sandy beaches aligned towards the SW swell.
Morfa Harlech, Gwynedd	SSSI/GCR Site	Comprises a major cuspate foreland with extensive sedimentation over historic times. It is significant for the relationship of the sand ridges to wave energy inputs, and is part of a suite of west coast beaches aligned to the SW swell.
Ynyslas Dyfed and Gwynedd	SSSI/GCR Site	Good example of a sand spit built upon a partly gravel base. Central area dominated by dunes. Significant for studies of estuarine sedimentation.
Oxwich Bay West Glamorgan	SSSI/GCR Site	Shows important relationships between dunes, beach and cliff. Close equilibrium between beach plan form and wave approach patterns.
Carmarthen Bay Dyfed and West Glamorgan	SSSI/GCR Site	Site includes connected series of beaches and spits fronting a rocky cliffed coast.

greater awareness of the need for dune management or stabilisation through afforestation and the less severe climate.

Sand dunes are of major conservation value, for both the natural habitats and important geomorphological features. Sand dunes managers, however, recognise the importance of a degree of wind erosion and stability, together with marine erosion of the seaward dunes, to maintain a full sequence of successful stages of habitats. In this context, change is a natural process with the stability imposed by many dune management techniques acting to preserve rare species rather than encouraging successful change.

Sand dune migration is principally a land management issue, rather than a planning and development issue. However, planners should be aware of the important role that sand dunes may have in providing natural coastal defences to low lying areas inland.

Maintenance of dune systems can, therefore, be an essential component of a coastal defence strategy. Planners should therefore, ensure that development within the dunes or on neighbouring stretches of coastline does not reduce their effectiveness through, for example, inappropriate mineral extraction or the disruption of coastal sediment

transport. The importance of these considerations has been highlighted in England and Wales through PPG20 Coastal Planning (DoE, 1992). It is also important that changes of land use in sand dune areas takes into account the potential for accelerated erosion if dune surfaces are left exposed to strong winds.

Chapter 11: References

Dargie T.C.D, 1993. Sand dune vegetation survey of Great Britain. Part 2: Scotland JNCC.
Dargie T.C.D., in press. Sand dune vegetation survey of Great Britain, Part 3: Wales. JNCC.
Department of the Environment 1992. PPG20 Coastal Planning. HMSO.
Doody, J.P., 1989. Conservation and development of the coastal dunes in Great Britain. In F van der Meulen, P D Jungerius and J H Visser (eds). Perspective in coastal dune management 53–68, SPB Academic Press.
Edlin H.L., 1976. The Culbin Sands. In J. Lenihan and W.W. Fletcher (eds.). Environment and Man Vol. 4 Reclamation; 1–13. Blackie.
Gresswell, R.K., 1953. Sandy shores in South Lancashire: the geomorphology of SouthWest Lancashire. Liverpool University Press.

Innes, J.B. and Tooley, M.J., 1991. The age and vegetational history of the Seflon Coast dunes. In D Atikinson and J Houston (eds) The Sand Dunes of the Sefton Coast, 35–40.

Jones, C.R., Houston, J.A. and Bateman, D., 1991. A history of human influence on the coastal landscape. In D Atkinson and J Houston (eds) The Sand Dunes of the Sefton Coast, 3–17.

Jungerius, P.D., 1989. Geomorphology, soils and dune management. In F van der Meulen, P D Jungerius and J H Visser (eds) Perspectives in coastal dune management 91–98, SPB Academic Press.

Jungerius, P.D., Verheggan, A.J.T. and Wiggers, A.J., 1981. The development of blow-outs in "De Blink" a coastal dune area near Noordwijkerhout, the Netherlands. Earth Surface Processes and Landforms, 6, 375–376.

Lamb, H.H., 1991. Historic storms of the North Sea, British Isles and Northwest Europe. Cambridge University Press.

Plater, A.J., Huddart, D., Innes, J.B., Pye, K., Smith, A.J. and Tooley, M.J. 1991. Coastal and sea-level changes. In D Atkinson and J Houston (eds) The Sand Dunes of the Sefton Coast, 23–34.

Radley, G.P., 1993. English coastal sand dunes and their vegetation: a national inventory. English Nature/JNCC.

Radley G.P., in press. Sand dune vegetation survey of Great Britain Part 1: England. JNCC.

Ranwell, D.S., 1975. In L Eugene (ed) Management of Saltmarsh and Coastal Dune Vegetation. Estuarine Research 2. Cronin Academic Press.

Ritchie, W., 1985. Scottish Beaches and dunes: a national survey for recreational management purposes. In: Gambling with the Shore. Proc. 9th Conference of Coastal Society, Atlantic City.

Ritchie W. and Mather A.S. 1984. The Beaches of Scotland. Countryside Commission for Scotland.

Stevens, C., 1992. The open coastline. In MG Barrett (ed) Coastal zone planning and management, 91–100 Thomas Telford.

Chapter 11: Suggested Reading

Doody, J.P., 1985. Focus on Nature No. 13. Sand dunes and their management. NCC Peterborough.

Carter, R.W.G., 1988. Coastal environments. Academic Press.

King, C.A.M., 1972. Beaches and coasts. Arnold Press.

12 The Occurrence and Significance of Erosion, Deposition and Flooding

The Occurrence

Erosion, deposition and flooding are natural phenomena associated with hillslopes, river sytems and the coast. The **hillslopes** are the site of soil erosion and deposition by water and, occasionally, wind; slope failure through landslides and debris flows also occur (these have been investigated previously in the Review of Landsliding in Great Britain; Geomorphological Services Ltd, 1986–87). Both supply sediment to the **river networks** where channel migration, flooding and deposition are the principal forms of geomorphological activity. The rivers reach the coastline in their **tidal estuaries** where sedimentation, flooding and the erosion of soft material by tidal currents are the dominant processes. On the **open coast** there are complex patterns of erosion and deposition, with tidal flooding in low lying areas.

The landscapes within which these processes occur are largely predetermined by topography, underlying geology and soils, the vegetation cover and land use. As has been emphasised throughout the Report these conditions control the erosion, deposition and flood behaviour of an area or region. For example, catchment characteristics are important controls on the nature and occurrence of floods (see Chapter 7). However, the geomorphological processes at work on the hills, rivers and coastline are not constant and do not take place in a static landscape. The occurrence of significant events is inexorably linked to the pattern of storms or severe weather conditions that are characteristic of Britain's maritime climate. Within this context, a number of key points are highly relevant to explaining the pattern of events:

- precipitation is usually highest in upland Britain with the result that most geomorphological energy is concentrated there. However, the uplands tend to be underlain by resistant rocks and, hence, much of the potential is often only realised in very large storms, or high intensity rainfall events.

- extreme rainfall conditions can occur throughout Britain, although outside the upland areas their occurrence is infrequent. As a consequence, major events are largely unexpected in many parts of lowland Britain and, hence, their occurrence can have severe consequences;

- the rate at which the flood discharges and, hence, flood height increases with larger return periods, varies across the country. For much of upland Britain, the 100 year flood event is expected to be around twice the annual flood (Figure 7.5). However, in southern England, east of a line from the Humber to the Exe, the 100 year flood event can be over 3 times the height of the annual flood. This pattern is largely a function of catchment character and has major implications for the size of defences needed to protect against floods of a given magnitude;

- stream power, a measure of the effectiveness of river flows in achieving erosion and deposition, tends to be as much as a thousand times greater in rivers in upland areas than in lowland areas;

- the energy arriving at the coast, through waves and tides, is considerably greater than that available inland. The coastline, therefore, can be the most dynamic environment in Britain, although the nature of the response is controlled by the strength and disposition of materials and landforms;

- although estuaries are perceived to be quiet backwaters, they do receive extremely high energy inputs as indicated by the semidiurnal input of over 1000Mm^3 of water into estuaries such as the Thames, Humber and Severn. However, these forces are controlled by the development of equilibrium morphology with changes occurring twice daily and, hence, almost imperceptibly. Thus, changes in estuaries are regular and slow, involving channel adjustments over extremely long time periods, in contrast to the open coast or rivers where change can be intermittent and rapid.

Beyond the seasonal pattern of climate, longer-term fluctuations in storminess or rainfall are important in determining the scale of geomorphological activity. For example, many of the major sand movement events described in Chapters 4 and 11 occurred during the period of colder and wetter climate known as the Little Ice Age. Over shorter timescales, periods of relatively wet or dry years can occur; these can have a profound influence on geomorphological activity, with many processes enhanced during major wet seasons. At the other extreme, sequences of dry years or drought conditions can significantly lower groundwater levels, reducing the likelihood of coastal cliff instability until sufficient recharge has occurred.

The geomorphological processes are linked. Flooding, for example, can initiate considerable erosion and deposition in hillslopes, rivers and on the coast. An important mechanism of linkage is through the transport of sediment from source areas to sinks where it is deposited in new landforms or as unconsolidated spreads on the sea or river beds. The sediment supplied from source areas can:

● affect the quality of water that is carried in a channel, influencing the rate of channel migration (Chapter 6);

● affect the quantity of water that can be carried in a channel influencing the flood risk or the storage capacity of a reservoir;

● protect slopes or coastal cliffs from further erosion;

● protect low lying coastal areas from flooding, through the build up of sand dunes, mudflats, beaches and salt marshes.

Sediment movement is generally an intermittent process with much deposited in temporary stores before being **delivered** to the next component of the system or out of the system altogether. Although much sediment transport is achieved by **extreme events**, especially for the movement of coarse material and boulders in rivers, it would be wrong to understate the importance of more frequent and less dramatic processes. Indeed, this is well illustrated by hillslope erosion where many of the examples listed in Chapter 3 have been initiated by unspectacular rainfall totals.

The physical systems : hillslopes, rivers and the coast, can be very responsive to changes in erosion or deposition patterns elsewhere in the landscape. For example, some river channels can readily adjust to changes in sediment supply by meander migration or the development of a braided network (Chapter 6). However, the **sensitivity** to change is not constant; it can vary considerably between landforms and systems.

The erosion, deposition and flood character of an area can, therefore, be seen to be influenced by:

● **linkages** between landforms within systems (the relationship between the site and its surroundings);

● the **sensitivity** of the system to change (i.e. the likelihood that events will result in changes in form).

A stable system is one in which the controlling resistances (i.e. strength of materials, equilibrium form, structural resistance) are such as to either prevent an event from having any effect or to be so arranged as to restore the system to its original state. For example, although wind and wave energy is greatest on the north and west coasts this does not correspond with rapid erosion and sediment transport. Consideration has to be given to whether the energy available can overcome the resistance of the system and whether the landforms are in equilibrium with the more extreme conditions.

The Effects of Man

The erosion, deposition and flooding character of many areas has been modified, to varying degrees, by the human occupancy of the landscape and land management practices. Amongst the more significant influences are those associated with changing socio-economic factors, including:

● **agricultural practice**; the general post-war intensification of agricultural production has been accompanied by removal of hedgerows, use of heavy vehicles, monoculture and winter cereal cultivation. All are believed to have contributed to an increase in soil erosion. Land drainage has had a significant effect on the speed at which both ground and surface water is carried to stream channels, increasing flood risk;

- **forestry**; after World War II, timber production was seen as of strategic importance, with the result that large areas of upland Britain have been afforested in the last 40 years or so. Between 1945 and 1983, 700,000ha was planted by the Forestry Commission representing the single largest land use change in Britain over this period (Acreman, 1985; Best, 1976). Around 60% of this land has been ploughed prior to planting (Taylor, 1970). It is believed that this has led to temporary increases in hillslope erosion and supply of sediment to stream channels, and more permanent changes in flood behaviour further down a catchment (see Chapter 5);

- **development**; the increasing demand for housing and employment opportunities by a growing population has led to an increased utilisation of floodplains and cliff–top locations. Frequently, these are viewed as desirable settings for expensive housing. Development has been accompanied by: an increase in the amount of impermeable surfaces within a catchment, increasing and accelerating runoff; reduction in floodplain storage following flood defence works; changes in flood behaviour following the construction of sewage and stormwater drainage systems; uncontrolled surface water discharge into slopes leading to increased erosion problems (see Table 12.1 and Chapters 3, 6 and 7);

- **river channelisation and flood defences** have often led to modifications to the patterns or erosion and deposition along a stretch of river, leading to river channel migration, and, in places, an increase in flood risk downstream. Flood embankments can prevent the overbank deposition of suspended sediments, leading to sedimentation problems downstream;

- **coastal defences** have frequently resulted in a disruption in the supply and transport of sediment around the coast. This, in places, has led to increased coastal erosion or flood risk, especially where natural defences such as sand dunes or beaches are deprived of a regular supply of sediment;

- **dredging operations** in some rivers and estuaries can lead to the movement of further sediment back into the deepened channel, resulting in a need for repeated dredging. This problem was recognised in the Thames estuary where the removal of $3Mm^3$ of dredged sediment each year to the outer estuary merely resulted in its immediate transfer back into the dredged channel (Inglis and Allen, 1957). By disposing of the spoil at sea, away from the outer estuary, the annual amount of dredging was reduced to around $0.25Mm^3$;

- **conservation**; the growing awareness of environmental issues since the 1940s has led to the designation of valued landscapes (e.g. National Parks, AONB's, National Scenic Areas) and conservation sites (e.g. SSSIs, National Nature Reserves, etc.). Development has been largely directed away from these areas, often increasing the pressure for urban expansion in floodplain areas or low lying coastal areas, e.g. in the Thames Valley. Conservation status has, of course, been used to restrict development in some areas of potential problems; it may, therefore, be the best land use option in vulnerable areas;

- **water supply**; the supply of cheap and clean water for the lowland centres of population from upland areas has transformed the behaviour of many rivers. Reservoir storage is essential to ensure a regular supply; these structures have also helped regulate flood flows. However, there have also been downstream implications for river channel form (Chapter 6).

In addition to modifying geomorphological activity, economic growth has resulted in an increase in the **damageability** associated with particular events by concentrating property, resources and services in vulnerable areas (see Chapter 7). For example, the expansion of properties and extensions in Datchet on the Thames floodplain has been well chronicled by the Flood Hazard Research Centre (Neal and Parker, 1988; see Chapter 7). One of the reasons that development has occurred in the flood prone area is that the settlement is entirely surrounded by

Table 12.1 A selection of the possible effects of development on aspects of erosion, deposition and flooding issues.

System	Potential Effects
Hillslope	• increase in runoff • increased infiltration through soakways etc and consequent effects on slope stability • creation of urban flood problems when rainfall intensity exceeds storm-drain capacity • modification of channel form can lead to local intensification of flood problems • bridges and culverts can be sites of temporary dams in flood events
River	• increase in peak flow and reduction in sediment supply can lead to channel change, especially increase in width and/or depth • effects on river corridor habitats and geomorphological features • reduction in floodplain storage and infiltration • land reclamation can lead to inter-tidal squeeze • flood embankments can cause loss of flood wave attenuation and lead to more severe flooding down stream (floodplain areas) or upstream (estuaries) • dredging may lead to an increase in sedimentation
Coast	• coastal defences, breakwaters etc. may disrupt sediment supply to beaches, dunes, mudflats etc. • mineral extraction from beaches and dunes or clay digging in estuaries may lead to decline in natural coastal defences and habitats • land reclamation or coastal defences can lead to loss of inter-tidal habitats • uncontrolled surface water drainage, leaking water pipes or sewer systems can lead to increased landsliding • excavation can lead to a reduction in slope stability • recreational pressures can lead to destabilisation of dune systems

Green Belt.

The Significance for the Environment

Throughout the report particular emphasis has been placed on how geomorphological processes are important in maintaining or shaping habitats or geological features. In general, erosion, deposition and flooding can:

• **maintain** habitats in river corridors and on the coast, through regular inundation or supply of sediment;

• **maintain** geological exposures or valued landforms along the coastline through continued erosion;

• **preparing** gravel bed rivers for spawning fish such as salmon;

• **create** valued landforms such as the fluvial geomorphological features associated with flash flooding or channel migration;

• **stimulate change** through promoting instability, ensuring that habitats evolve through natural successions, rather than remaining static.

Erosion, deposition and flooding processes have created extensive spreads of river valley gravels and marine aggregates. Extraction of these resources yields around 100M tonnes a year at an estimated value of £500M, (Table 12.2). These deposits are largely relics, having been laid down many thousands of years ago under different environmental conditions. Contemporary replenishment is believed to be minimal and the resource finite.

The importance of erosion and deposition in minimising the impact of extreme events should not be underestimated. For example, many coastal landforms offer a degree of protection against coastal flooding. **Sand dunes** serve as a natural barrier against high water levels; this has long been an effective coastal defence for many communities around the coast. **Beaches** and **shingle ridges** absorb as much as 90% of the wave energy arriving at the coast by continuously adjusting their form (Brampton 1992), providing an important component of sea defences where they front embankments or sea walls. **Saltmarshes** and **mudflats** are also effective in dissipating wave energy. All these landforms are dependent upon a continued supply of sediment to maintain their form; disruption of sediment transport can, therefore, lead to an increase in the degree of risk of erosion behind the depleted landforms.

Table 12.2 Sand and gravel production, 1991 and estimated values.

Area	Extraction (Thousands of Tonnes)		
	Land Won	Marine	Estimated Value (EM)
East Anglia	7,288	*	35.7
East Midlands	12,683		59.6
North	3,017	*	26.2
North West	4,243	241	19.4
South East	24,282	9,035	174.8
South West	6,045	504	27.9
West Midlands	10,698		47.6
Yorks & Humberside	4,960	*	22.8
Wales	1,836	1,603	18.3
Scotland	12,226		52.4
Total	85,479	12,439	484.7

* no data available

Sources: Business Monitor PA1007, 1991; BACMI Statistical Yearbook, 1992; Estimated values based on approach developed by Arup Economics and Planning, 1993.

The importance of these natural processes is now widely appreciated by coastal defence operating authorities and conservation groups. Indeed, this view is central to English Nature's Campaign for a Living Coast (English Nature, 1992):

"English Nature will seek to halt and reverse the loss of coastal habitats and natural features resulting from coastal squeeze and from the disruption of natural sedimentary systems. We shall try to establish a principle that new or replacement sea defence, coast protection or similar works should not exacerbate coastal squeeze or disruption of systems and **should reverse these whenever possible**, so as to maintain habitats and natural features **at least equivalent to their present distribution** (1992) and in a sustainable condition." (English Nature, 1992).

It would be wrong, however, to imply that natural processes are always of benefit to conservation interests. Historic monuments on the coast are vulnerable to erosion, deposition and flooding. Clearly, the management of these sites may involve some form of protection and could be in direct conflict with other conservation interests.

Erosion, Deposition and Flooding as a Hazard

These are natural processes; they only become hazards when development or land use encroaches into vulnerable areas. It is clear that, despite the post-war structural responses, that the processes have increasingly imposed themselves upon many communities, creating frequently unexpected problems to homeowners and businesses who were largely unaware of the risks that could be anticipated. This is readily apparent on the broad floodplains of Britain's major rivers, on the soft rock cliffs of eastern and southern England, and the coastal lowlands of north Wales and England.

The problems associated with these processes are generally similar: loss of agricultural productivity, damage to property, services and infrastructure, the need for regular maintenance to ensure unhindered use of waterways; and occasionally death or injury. It has been estimated from the historical record that the average level of damage or maintenance and defence needs associated with these processes probably exceeds £300M per year (Table 12.3); these costs are, of course, spread through many levels of the economy, from individuals to industry, local authorities to national government

155

Table 12.3 An indication of the general order of costs per year arising as a consequence of erosion, deposition and flooding.

Hillslopes	Estimated Annual Cost £	Source
	Unknown	
Rivers ● NRA maintenance ● British Waterways dredging	£42M £3M	NRA Corporate Plan (1993/94) British Waterways News (1993)
Coast and Estuaries ● NRA maintenance ● dredged spoil	£15M £43M	NRA Corporate Plan (1993/94) Table 8.2; assumed £1 per tonne for disposal
Flood and Coastal Defences ● NRA ● Local Authorities	£115M £65M	NRA Corporate Plan (1993/94) DoE Statistics
Damages ● impact of events in historical record	£50M	Based on values in Table 2.2
Estimated Total	£333M	

Note: Investment in capital works is justified on the basis that its cost is less than the damage avoided, so that without this expenditure the overall cost would be significantly higher.

and include:

(i) **direct damages** caused by the effects of erosion and deposition, or the physical contact of floodwater with properties and their contents;

(ii) **indirect damages** arising as a consequence of direct damage, including: traffic disruption, loss of production, evacuation costs, etc.;

(iii) **intangible damages** ranging from anxiety and stress to ill health related to the general inconvenience caused by the event.

However, it is the geographical extent and the intensity of damage, disruption or personal losses that set some processes apart. Flooding – flash floods in upland areas, lowland river floods and tidal floods – is the most dramatic and costly problem for society. Examples of particularly distressing and costly events can be found in the historical record for many parts of Britain; some of the worst include:

● the flash floods in the Lynmouth area of Devon on 15 August 1952 when 34 were killed and £9M of damage (at 1952 prices) was caused;

● the lowland floods of March 1947 which affected rivers throughout South Wales and much of England. The resulting damages

were probably in excess of £500M (at current prices);

● the east coast floods of 31 January 1953 when 300 died and damages were an estimated £900M (at current prices);

● the Severnside coast floods of 1606, when about 2000 people drowned as sea defences were overtopped;

By contrast with the spatially extensive problems associated with flooding, other processes tend to create **site specific** or localised difficulties. Even so, they can still pose a significant threat to construction and development or lead to high maintenance costs to alleviate the effects of the processes. The erosion of coastal cliffs and sedimentation in rivers and estuaries can be a significance constraint to human activity, as illustrated by:

● the major coastal landslide at Holbeck Hall, Scarborough in June 1993 which is likely to have resulted in excess of £3M of damage and repair works;

● the annual maintenance dredging costs in excess of £1M incurred at Harwich and Liverpool. At Kings Lynn, the approaches have to be resurveyed every two weeks with navigation buoys repositioned up to 100 times a year.

Deposition within river channels can lead to serious maintenance and operational problems. In 1993, for example, British Waterways spent over £3M on dredging, involving the removal of 300,000 tonnes of material. These operations are necessary to ensure that British Waterways fulfils its statutory obligations, but the need to dispose of the material on land can lead to conflict with local planning authorities and, in England and Wales, the NRA.

Other processes such as hillslope erosion, wind erosion and channel migration can lead to notable problems for affected landowners and can lead to difficulties where infrastructure and services cross vulnerable areas. The implications of these problems are easy to dismiss as trivial; the following examples should serve to demonstrate that they can lead to serious problems:

- soil erosion and mudflood problems in the South Downs during October 1987 probably resulted in £0.75M of damage, especially in and around Rottingdean;

- channel scour and erosion around the piers of a railway bridge at Glanrhyd, Dyfed led to the bridge collapsing under the weight of a train in October 1987; four people died;

- the Culbin Sands disaster of 1694 and following years led to over 20–30km^2 of fertile farmlands, near Findhorn on the Moray Firth, being buried by up to 30m of loose sand. The estimated damages were probably the equivalent of £25M, at present prices.

Many relatively minor erosion events can also achieve importance because they supply sediment to rivers or the coastal zone. Sediment supply occurs in areas of **hillslope erosion** and where **river channel migration** cuts through areas of stored sediments resulting from past phases of erosion under different climatic conditions or from an extreme flood in the recent past (e.g. spreads of glacial deposits or floodplain alluvium). Once in the channel, the sediment size is important in determining how far it is carried before being temporarily stored in features such as point bars or as spreads on the river bed. In short rivers the suspended load may reach the estuary in a single flood, but coarser sediments may become incorporated in the floodplain. Deposition within the channel can, of course, reduce its capacity and lead to flood problems, as reported for the River Spey in Grampian and the Findhorn in Highland. It can also lead to navigability problems, scouring of

bridges and other engineering structures, sedimentation and loss of reservoir capacity and a decline in water quality. Deposition around the coast can help maintain natural coastal defences, such as sand dunes and beaches, which protect vulnerable areas from flooding or cliff erosion.

Potentially vulnerable areas can be readily identified at a general level; the most important sources are likely to be catchment management plans, shoreline management plans and S.105 surveys. The accompanying 1:625,000 thematic maps are a general summary of the national distribution of areas where particular problems may be anticipated. These include:

- erodible soils
- unstable river sections
- floodplains
- coastal lowlands
- eroding coastal cliffs
- sand dunes

Although areas can be defined there are major problems in evaluating the **potential** for hazardous events to occur. Damaging events are often associated with very large return period storms (the Lynmouth floods of 1952 had an estimated return period of thousands of years) and are generally the product of a combination of circumstances (e.g. antecedent rainfall conditions, groundwater levels, land management practice). Further complications arise when the degree of existing protection afforded by defence works are considered. These can reduce the hazard not eliminate it, as they are designed to protect against a certain return period event. Events greater than that magnitude can cause a failure in the defences.

Return period statistics are, however, notoriously unreliable, especially when derived from very short periods of records, where new events can lead to a significant modification in the calculated return period. For example, the Truro floods of January 1988 initially had an estimated return period of 350 years, using the procedures recommended in the Flood Studies Report (NERC, 1975); reappraisal after the floods of 1988 suggested a return period of 50 years for the January event (see Chapter 5). This **probabilistic approach** regards individual events as a random part of a natural series of events of varying magnitude and frequency. Thus, the flood which is expected to be equalled or exceeded, on average, every 100 years, has a return period of 100 years. This event could occur any year; indeed, the probability of it occurring in a 30 year period is around 25% (Table 2.1). This has obvious significance to the degree of risk from extreme events. Even where defences provide a 1 in 100 year return period level of protection there remains a 45% chance of a damaging event in the 60 year lifetime of a building.

A different perspective on hazard potential is provided by considering the degree of risk in terms of the standard of protection which exists in an area. MAFF, for example, have recently set out indicative standards of protection in terms of flood return periods for five subjectively expressed "current land use bands" (Table 12.4). These standards are intended to help establish the range of options to be considered for particular problems. They do not represent an entitlement or a minimum level to be aimed at. In many instances, the standard of protection can give a clearer indication of the hazard potential than the use of return period statistics. Potentially vulnerable areas can be defined in terms of the land use band which has the closest indicative standard of protection. This approach could highlight the level of improvements to the defences that would need to be incorporated into the design of new developments.

Notable difficulties also arise in the prediction of erosion rates, especially on coastal cliffs. As noted in Chapter 10, cliff recession is an episodic process; there may be little or no erosion for a long period, followed by sudden rapid recession in response to a large storm. The "rate" of recession may, therefore, depend on the timescale chosen. Furthermore, the main causal factors such as wave undercutting are stochastic, i.e. inherently unpredictable. This uncertainty creates problems for coastal managers when attempting to define the degree of risk to cliff top property, either for identification of "set-back" lines by planners or the choice of erosion control technique by engineers. In this context, it is worth noting that MAFF have recently established a programme of research to develop methods of predicting soft cliff recession rates suitable for both detailed design and strategic planning.

Future Considerations: Erosion, Deposition and Flooding in a warmer Britain

There has been much debate in recent years of the global warming/rising sea level issue. The Intergovernmental Panel on Climate Changes has identified the extent to which rising sea levels are in evidence around the world, and the extent to which further rises may be expected (Houghton et al 1990). The Second World Climate Conference (Jager and Ferguson, 1991) reached similar conclusions, which in the case of the British Isles suggest that there could be arise of between 50 and 70cm over the next 100 years. However, it is clear that this rise would not be the same in all parts of Britain since long-term vertical land movements are still taking place in some areas, and the more northerly parts (e.g. the coastal margins of the Highlands of Scotland) may continue to see a relative drop in sea level (Figures 12.1 and 12.2).

As yet, British tidal gauge records show no clear evidence of an acceleration in the rate of sea level rise (Woodworth, 1990; Woodworth et al, 1991). However, even if there is no acceleration, mean sea level is predicted to rise by as much as 10cm over the next 20 years on parts of the south coast (Bray et al, 1992). Allowances given by MAFF (November 1991; see also DoE Circular 30/92) for the design or adaption of coastal defences with an effective life beyond 2030 range from 6mm per year (south east and southern England) to 4mm per year (north west and north east England) and 5mm per year (the remainder of England and Wales).

In terms of climate, the "best estimate" of change to the year 2050 (Hulme et al, 1993) indicates that there could be notable modifications to the erosion, deposition and flood character of many areas. The most important influences are likely to be:

● higher winter rainfall;
● increased likelihood of summer droughts;
● increased summer storm activity.

A wide variety of studies have indicated the potential for changes in the nature and rate of physical processes in different physical environments (e.g. Doornkamp 1990; Parry et al., 1991, Boardman 1993; Bevan 1993; Shenna 1993). However, any attempts to define the likely changes in magnitude, frequency and impact of future events are constrained by a number of problems:

● the changes are likely to be extremely varied in character, reflecting the varying sensitivities of different catchments or coastal systems;

● events are frequently the consequence of the interaction between a range of factors of which climate change is merely one set of controls;

● many problems are a reflection of local conditions which are very difficult to

Table 12.4 Indicative standards of protection (after MAFF, 1993).

Current Land Use	Indicative Standard of Protection (Return Period in Years)	
	Tidal	Non–Tidal
High density urban containing significant amount of both residential and non–residential property.	200	100
Medium density urban. Lower density than above, may also include some agricultural land.	150	75
Low density or rural communities with limited number of properties at risk. Highly productive agricultural land.	50	25
Generally arable farming with isolated properties. Medium productivity agricultural land.	20	10
Predominantly extensive grass with very few properties at risk. Low productivity agricultural land.	5	1

Note: This table should not be used in isolation from the relevant Project Appraisal Guidance notes (MAFF, 1993).

predict at a general level.

There appears to be general agreement that extreme climatic events are likely to be more frequent over the next century; this could lead to a significant increase in the magnitude of impact arising from erosion, deposition and flooding events. An indication of the potential severity of events is provided in the historical record of **great storms** that have been reported throughout the last 500 years (Table 12.5; Lamb 1991).

The historical record also provides an opportunity for developing scenarios for change in the erosion,

Figure 12.1 Estimated current rates of crustal movement (mm/yr) in Great Britain (after Shennon, 1989).

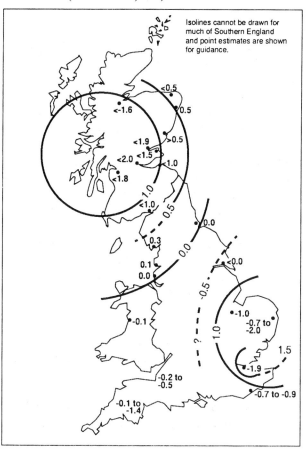

Figure 12.2 Recent sea level changes (mm/yr) around Great Britain (after Carter, 1989).

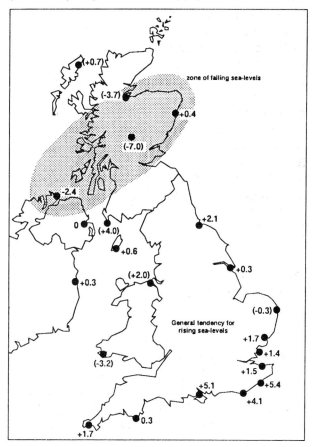

159

Table 12.5 Examples of some of the severest storms to have affected Great Britain over the last 500 years (after Lamb, 1991).

Date	Storm Severity* Index	Comment
7–8 December 1703	9000	Reported to have been the severest storm on record; 8000 died. Wind speeds up to 150 knots. Short (1749) wrote "England lost more ships in this storm than ever were lost in any encounter with the enemy". Extensive flooding; tides at Bristol were 2.5m higher than previously recorded.
6–7 January 1839	8000	400 dead in British Isles; insurance losses estimated at £1–5M (1839 prices). 115 died in Liverpool. Winds up to 100 knots.
16 October 1987	8000	Enormous damage to buildings, trees, electricity and telephone lines; 18 dead, estimated insurance losses of £1000M (1988 prices).
14–16 October 1886	7000	Heavy gale, accompanied by rain and flooding in England and Wales; bridges washed away.
31 January to 1 February 1953	6000	Great storm with gusts up to 109 knots; accompanied by tidal surge on the east coast of around 2.5m at King's Lynn. 350 dead in England.
3 January 1976	6000	24 dead; enormous numbers of trees blown down; insurance cost estimated at £126M (1976 prices).
1570	5000	Breckland storms led to movement of great quantities of sand across East Anglia, creating the "Sand Floude" that overwhelmed Santon Downham.
25–27 January 1884	5000	Very widespread strong gales; great damage to trees and woodlands. More than 1M blown down trees on one estate in Galloway.
17–19 November 1893	5000	Prolonged and strong gale, leading to extensive snow falls; drifts up to 1.5m deep in Surrey. Great destruction of woodlands throughout Britain.

Note:
*The storm severity index is the product of the greatest wind speed, the area affected and the overall duration. Lamb (1991) has classified historical storms into the following grades of severity according to this index:

Class I;   5000 and over   Class IV;   600–600
Class II;   1800–4000   Class V;   750–250
Class III;   1600–1700   Class VI;   less than 100

deposition and flood character in response to variations in climate. Indeed, it is possible to match climatic records over the last 300–400 years with documentary accounts for major events and early map sources. Within this period there have been notable variations in storminess, not only from year to year but also over decades and centuries. Figure 12.4 presents the number of storms of different severity class since 1570 and reveals that a number of marked periods of increased storminess: prior to 1650; 1880–1900; and since 1950 (Lamb 1991).

The storminess prior to 1650 is widely believed to reflect the period of colder, wetter climate known as the "Little Ice Age". This period was characterised by frequent severe winters, reduced run–off (Thom and Ledger, 1976, suggest that runoff was 89% of present levels) and the occurrence of surface winds of strengths unparalleled in this century. Indeed, most of the major wind blown sand events are from this period

(e.g. the Culbin Sands disaster of 1694 and the Breckland storms of between 1570-1588; Table 11.1).

The end of the Little Ice Age was marked by a much wetter, more extreme and variable climate which may offer an analogue to the current phase of atmospheric warming (Newson and Lewin, 1991). This period from 1700-1850 has been associated with:

● marked increases in autumn and winter floods, as identified on the River Wear in Durham (Archer, 1987);

● major river channel changes in both the uplands of northern England and in floodplain locations (Macklin et al, 1991);

● an increase in the reported incidence of major coastal landslides in southern

160

England (e.g. the 1810 Landslip on the Isle of Wight, the Great Bindon landslip of 1839; Jones and Lee, 1995);

- an increase in debris flow activity in the Highlands of Scotland (Innes, 1983).

These patterns of changing erosion, deposition and flood character have to be superimposed on the general trend of increasing frequency of damaging events identified in Figure 2.24 which highlighted the effects of:

- the rapid spread of development into vulnerable locations throughout the 19th and 20th centuries;

- the improvements to flood warning and defences since 1950 which have resulted in a reduction of risk in protection areas.

However, if global warming and sea level rise are to generate more extreme climatic events, then this will have inevitable consequences on the estimated return periods for potentially damaging events. In some instances this may result in the present 1 in 100 year flood event becoming as frequent as 1 in 10 years (Pethick and Burd, 1993), with obvious consequences for the standards of service provided by flood and coastal defence schemes.

The possible effects of global warming and sea level rise are likely to heighten catchment and coastal problems which are already experienced, albeit on an infrequent basis. The possible scenarios raise not so much the issue of the specific policies that are required to meet it as the need to have in place appropriate management framework to effectively cope with such problems as and when they arise.

Sustainable Development: Management of Erosion, Deposition and Flooding Processes

The UK strategy for achieving sustainable development (Secretary of State for the Environment and Others, 1994) is based on a number of specific principles:

- decisions should be based on the best possible scientific information and analysis of risks;

- where there is uncertainty, and potentially serious risks exist, precautionary action may be necessary;

- ecological impacts must be considered, particularly where resources are non-renewable or effects may be irreversible;

- cost implications should be brought directly to the people responsible – the "polluter pays" principle.

A key theme throughout this Report has been how flooding and coastal erosion can present significant constraints to land use and development, imposing very high costs on society through damage to property, services and infrastructure, emergency relief and the other resources devoted to management of the problems. Potential problems associated with other processes (e.g. hillslope erosion, channel instability and sedimentation in rivers and estuaries) should not be dismissed as trivial matters; they can impose significant costs on specific sectors of the economy (e.g. water companies, navigation and harbour authorities) and increase the risk of potentially damaging flood events. Where the risks to use and development are at an unacceptable level, precautionary measures may be necessary to control erosion, deposition and flooding processes. Precautionary measures needed to reduce the levels of risk may, however, have significant and irreversible effects on some environmental resources. Judgements clearly have to be made about the weight to be put on these factors in particular cases. Sometimes the environmental costs may have to be accepted as the price of economic development, but on other occasions resources may be so valuable that they have to be protected from the potential effects of development.

The issues arising from erosion, deposition and flooding processes can be related to **development, conservation, recreation** and **mineral supply** (Table 12.6). However, the broad scale operation of these natural processes dictate that it is generally inappropriate to address these issues on a site specific basis; strategic considerations are necessary on a catchment or coastal system–wide basis. Indeed, the complex operation of processes within these natural systems present a number of unique challenges to decision–makers:

- erosion, deposition and flooding problems are often linked and should not be treated in isolation. Flooding, for example, can initiate considerable erosion and deposition

Table 12.6 Key issues associated with erosion, deposition and flooding processes.

Development

- the impact of erosion, deposition and flooding processes on development and the need for remedial and defence works;
- the effects of development on the operation of physical processes within a catchment or along the coast;
- the effects of development on mineral and conservation resources.

Conservation

- the importance of physical processes in creating and maintaining conservation features;
- the effects of conservation policies on diverting development into vulnerable areas away from conservation resources.

Recreation

- the importance of physical processes in creating and maintaining recreation resources such as beaches and sand dunes;
- the effects of recreational facilities on the stability of features such as sand dunes.

Minerals

- the occurrence and significance of mineral resources such as aggregates for construction and coastal defence, and building stone;
- the effects of mineral extraction on the operation of physical processes within a catchment or along the coast;
- the effects of mineral extraction on coastal, marine and riverine resources;
- the conservation opportunities provided by mineral workings.

on hillslopes, rivers and on the coast. Flood problems are often exacerbated by erosion and deposition;

- development may lead to significant changes elsewhere within a system. These changes can affect the level of risk elsewhere or lead to the degradation of both natural landforms and the habitats which they support;

- the cumulative effects of development on a system may take many years to become apparent. Indeed, past responses to the threat of erosion and flooding have led to many long term problems facing river and coastal managers.

It is important that management decisions are based on the best possible understanding of the physical environment and how to exploit, manage and protect it. In this context, the management issues presented in Table 12.6 can be addressed in terms of three key considerations which are central to the sustainable development and use of river catchments and coastal systems:

- the **risks** associated with geomorphological processes, as highlighted in Chapters 3–11;

- the **sediment budget** of a catchment or coastal system which can be a major factor

in ensuring the sustainability of natural and engineered defences, navigation uses, port and harbour operations, conservation sites and recreation areas, and operations such as aggregate extraction.

- the **sensitivity** of the physical environment to natural or man–induced change. This can be an important factor in assessing the potential effects of development along a river or coastline.

These key considerations are relevant at all stages of the decision making process, from regional or strategic level to the determination of site specific proposals. In the complementary report **"Investigation and Management"** a hierarchical model was set out which indicated the investigation approach suited to different stages of the decision-making process; from general assessment to site reconnaissance and detailed assessment. (Rendel Geotechnics, 1995). As the information required for these investigations changes from a general awareness of the coastal environment to the need for site specific information, the questions that should be considered remain broadly the same (Figure 12.3). The significance of some questions and the extent to which they will need to be addressed will, however, vary according to the nature of the environment and the extent to which it is subject to pressure for development.

Figure 12.3 Elements of earth science investigations.

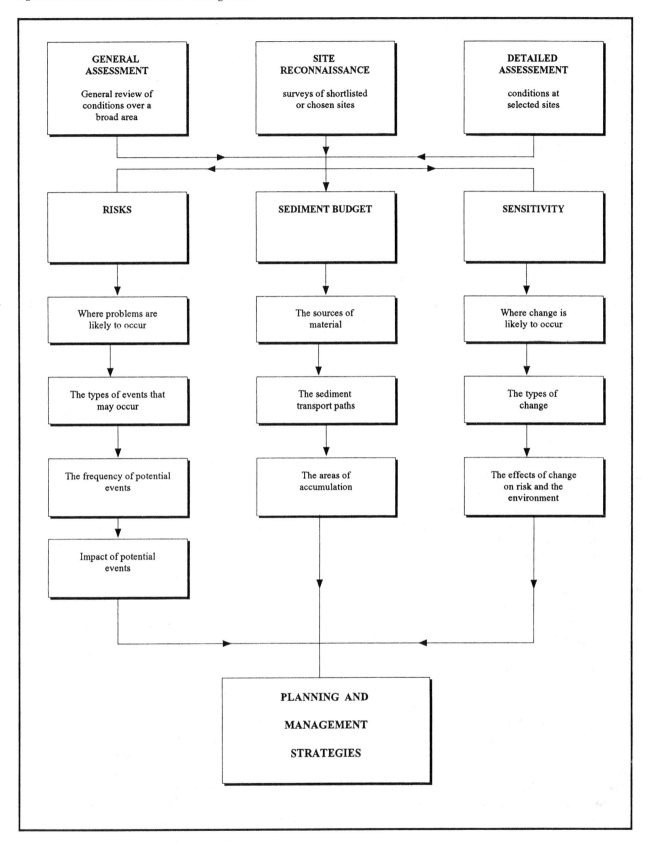

Figure 12.4 Storms of different severity class since 1570 (after Lamb, 1991).

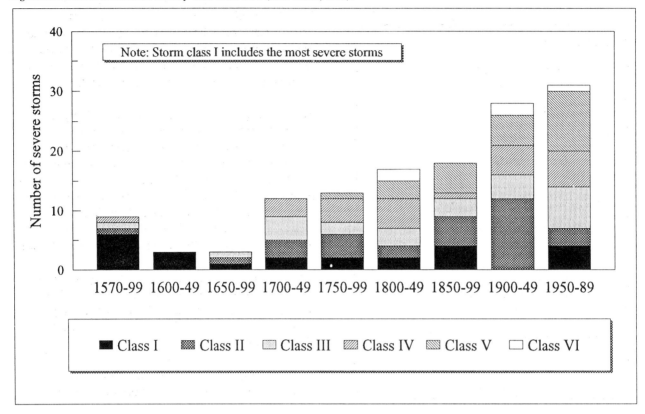

A wide variety of responses are available for managing physical processes on hillslopes, rivers and the coast (Table 12.7). The most appropriate option will depend on the nature of the problem, the level of acceptable risk, the availability of resources and the statutory powers available to interested bodies or authorities to tackle the problems. In most cases the response will be complex, involving a variety of measures adopted to different locations within a catchment or coastal system.

The complexity of the inter–relationships between the three processes and management practices, and their combined effects on a variety of land use interests highlight the need for a coordinated approach to decision making. This would generally involve strategic or regional level awareness of how the effects of an activity or proposed development at one location can be transmitted over a wider area and impose a variety of problems on third party interests. One response to this need has been the development of system–wide strategic plans by the NRA (Catchment Management plans) and Coastal Defence Groups comprising authorities with coastal defence responsibilities along a length of coastline (Shoreline Management Plans).

Erosion, Deposition and Flooding as a Planning Issue

Prior to the mid 1980's, local planning authorities frequently viewed natural hazards such as erosion and flooding as technical problems that the landowner and developer needs to overcome or the responsibility of coast protection authorities and drainage authorities; they were not seen to be land use planning issues. Since the mid 1980's there has been a notable change in perception about the way in which problems are managed. These changes reflect a growing appreciation that the past approach was not in the public interest:

•   development in vulnerable locations can lead to demands for expensive publicly funded defence works;

•   the possible adverse effects of development on the level of erosion or flood risk elsewhere;

•   defence works can have significant adverse effects on the interests of other users of rivers or the coastal zone;

•   defence works can encourage further development in vulnerable areas, increasing the potential for greater losses when extreme events occur.

164

Table 12.7 Management responses.

```
┌─────────────────────────────────────────────────┐
│  Acceptance of the Risk                           │
│                                                   │
│    •      maintenance and repair                  │
│    •      emergency planning                      │
│    •      insurance                               │
│    •      litigation                              │
│                                                   │
│  Avoiding Vulnerable Areas                        │
│                                                   │
│    •      land use planning                       │
│    •      managed retreat                         │
│                                                   │
│  Reducing the Likelihood of Potentially Damaging Events │
│                                                   │
│    •      regular reservoir inspection            │
│    •      hillslope management                    │
│    •      runoff management                       │
│    •      floodplain management                   │
│    •      river channel management                │
│    •      coastal cliff management                │
│    •      foreshore management                    │
│                                                   │
│  Protecting Against Potentially Damaging Events   │
│                                                   │
│    •      early warning                           │
│    •      defence schemes                         │
│    •      building modifications                  │
└─────────────────────────────────────────────────┘
```

The change in attitude also reflects concern about the possible effects of global warming and sea level rise and, at a local level, the effects of recent major hazard events such as: the North Wales floods of February 1990; the Tayside floods of 1990 and 1993; and the Holbeck Hall coastal landslide of June 1993.

The Government has advised planning authorities in England and Wales that it is the purpose of the planning system to "regulate the development and use of land in the public interest" and that planners need to take into account "whether the proposal would unacceptably affect amenities and the existing use of land and buildings which ought to be protected in the public interest" (PPG1, DoE 1992a; PPG12, DoE 1992b). Clearly development in vulnerable areas is not in the public interest if appropriate preventative or precautionary measures have not been taken, or if they lead to significant adverse effects on the environment or other interests. The potential impacts of development proposals on the public interest are clearly land use planning issues, especially when public funds are sought later to protect against natural hazards.

In certain areas physical processes, in particular river and coastal flooding and coastal erosion, can impinge **directly** on the land use planning functions of local planning authorities. This will include situations where proposals for new development or redevelopment take place in areas which may be at risk, or where proposals to protect existing areas of land from physical processes can have an impact on other planning objectives.

Erosion, deposition and flooding can also generate **indirect** land use issues. Bank erosion, together with sediment delivered to watercourses from hillslopes, can lead to sedimentation problems; the response, maintenance dredging, can cause land use issues when the material has to be disposed of on land or if the operations threaten to damage conservation interests. Conversely, the cessation of dredging following the closure of port and harbour facilities may lead to the degradation of the riverbank environment in city centres (e.g. on the Clyde) or cause an increase in flood risk. These matters can clearly be considered as planning issues and may need to be addressed by local planning authorities.

The planning system clearly has an important role in hazard management, most notably through:

• avoiding unsuitable areas or specifying restrictions on housing occupancy in risk areas;

• ensuring that precautions are taken to prevent runoff from new developments increasing flood risk;

• ensuring that development does not adversely affect floodplain storage and, hence, increase flood risk;

• facilitating the disposal of dredged material from navigable waterways by identifying suitable disposal sites;

• ensuring that development does not affect coastal stability or lead to an increase in coastal erosion;

• ensuring that flood and coastal defence works do not have significant adverse effects on other interests;

• supporting environmental management objectives by ensuring that development does not lead to a decline in value of conservation sites dependent on the continued operation of physical processes.

The effectiveness of the planning system can be enhanced significantly by ensuring that:

(i)   erosion, deposition and flooding issues are considered at a strategic or regional level, especially in areas where the effects of development may lead to an increase in the degree of risk or cause significant environmental effects elsewhere in the catchment or coastal system;

(ii)   local planning authorities liaise with catchment and shoreline managers in the preparation of development plans and policies which take full account of the need to minimise risks. In this context, it is important to stress that:

- construction of defences often leads to increased pressure for development in what is now perceived to be a safe area;

- flood and coastal defences can only reduce the degree of risk, not eliminate it;

- increased investment and density of development behind defences may lead to higher losses when, inevitably, larger floods occur.

These issues are examined in further detail in the complementary report prepared for this project, entitled "**Investigation and Management**" (Rendel Geotechnics, 1995).

Conclusions and Recommendations

In this Report it has been shown how erosion, deposition and flooding can represent significant problems for land use planning and development, most notably through:

- the impact of the processes on property, infrastructure and services;

- the effects of development on the degree of risk elsewhere;

- the conflicts generated by the selection of management strategies.

In areas where erosion, deposition and flooding are likely to impose constraints to development and land use, decision makers will need to consider identifying those areas where particular

consideration should be given to these issues. They may also need to be aware of the type of problems that may occur, how frequently that damaging events may take place and whether they can be satisfactorily overcome by proposed developments or uses.

In general, planners and developers require guidance on:

- where erosion, deposition and flooding issues need to be considered;

- the nature of the hazards in the areas and the general level or risk to existing land uses and development;

- whether potentially damaging events can be expected during the normal lifetime of particular types of development;

- the types of uses or development that, in general terms, may be best suited to those areas;

- whether development or land use can have adverse effects on other interests in the river or coastal environments.

This requires access to reliable technical information and although there is a considerable volume of available data on erosion, deposition and flooding, little is directly suitable for planners insofar as it could be used without re-interpretation. There is, therefore, a clear need to improve the availability of information that is needed to support forward planning and development control. This should involve considering the following recommendations.

Basic Data Collection

1.   Decision makers should be aware of the level of damage and disruption that can occur as a result of erosion, deposition and flooding. At present this information remains largely inaccessible and in a wide range of formats. **It is recommended that a systematic approach to recording the impact of erosion, deposition and flooding processes should be developed.** This could involve:

1a.   Developing a standard approach for use by highways authorities, drainage authorities, coast protection authorities and other bodies;

1b. Encouraging navigation and harbour authorities and water companies to record maintenance dredging requirements and make the data available for wider usage;

1c. Establishing a mechanism whereby the event records can be collated and made available to planners and developers.

2. It is particularly important for decision makers to be aware of the historical incidents of damaging events in an area as this can provide an indication of the nature and scale of extreme events. For most major rivers and coastlines these historical studies have been carried out for flood and coastal defence purposes. This data source has, however, been under-used and it is recommended, therefore, **that local planning authorities are encouraged to seek access to these reviews of historical sources to improve their awareness of extreme events that have occurred within their area.**

3. Historical map sources and aerial photographs can provide a general indication of the cumulative effects of erosion and deposition between survey periods, as well as the variability of the rates of recession and accretion. **However, use of these sources is not straightforward and it is recommended that guidance is provided to local planning authorities and developers.** This could address:

–   methods for assessing confidence limits to estimated recession and accretion rates;

–   the use of geomorphological techniques to support the historical appraisal;

–   factors which may invalidate predictions from historic data, e.g. environmental change, geological variability.

4. Although much information is currently available on the distribution of problems associated with erosion, deposition and flooding, little is known of the benefits to the environment. This is particularly apparent when trying to identify those landforms, habitats and resources that are dependent on the continued operation of the processes. It is recommended that **those bodies with responsibility for various aspects of resource management identify those features which could be affected by a disruption to the natural patterns of erosion, deposition and flooding; this information should be made available to all those involved in the planning and development process.**

5. In many instances the identification of vulnerable areas can be achieved from readily available sources. However, there are a number of notable exceptions. **Considerations should therefore be given to undertaking systematic surveys of the areas affected by, or likely to be affected, by the following problems:**

–   mudfloods associated with hillslope erosion events;
–   hillslope erosion in Scotland;
–   flood risk areas in Scotland;
–   watercourses and estuaries with notable sedimentation problems;
–   mobile sand dune areas;
–   areas of river channel instability and bank erosion.

Data Presentation

6. It is important that the considerable volume of information on erosion, deposition and flood hazards is made readily accessible to potential users in the planning and development communities. It is recommended, therefore, **that organisations with an interest or responsibility for data collection and presentation be encouraged to develop standard approaches to preparing summary hazard maps and non-technical reports which identify:**

–   areas where problems are likely to occur;
–   the nature of the hazards and the potential risks in these areas;
–   the level of protection provided by existing or proposed defences;

– recent and current rates of erosion and deposition.

7. The sediment budget of a catchment or coastal system is likely to be a central consideration for ensuring that the objectives of sustainable development are met. It is recommended, therefore, that **operating authorities with a responsibility in catchment and shoreline management should make this information available to planners and developers.**

8. The effects of a development on the level of risk and the environment elsewhere within a catchment or coastal system is an important consideration for land use planners. However, there are currently few sources of information about the sensitivity of different physical systems to change that can be readily used by planners and developers. It is recommended, therefore, that **techniques are developed for evaluating the sensitivity of an area to the effects of development, which are suited to a range of environments.** This could involve:

    – development of empirical or probabilistic models for use over relatively small areas;
    – development of broad-brush mapping techniques for use over catchments or coastal systems.

Data Storage and Retrieval

9. The historical record database developed as part of this study provides a ready source of information concerning the pattern of significant events over the last 200–300 years or so. Whilst the database has significant limitations, it does provide a valuable starting point for research into the frequency of erosion, deposition and flooding events. It is recommended, therefore, **that the database is made available to academic researchers and that resources are provided to support the integration of the records with related datasets.**

10. The historical records and thematic maps prepared as part of this study provide the basis for developing an integrated model of the temporal and spatial aspects of the erosion, deposition an flooding character of Great Britain. **Resources should be provided to encourage the development of a computerised Geographical Information System which incorporates the data collected as part of the study and from other sources, to define and model the variations in erosion, deposition and flood character across the country.**

Monitoring Environmental Change

11. Global warming and sea level rise could have significant effects on the erosion, deposition and flood character of Great Britain, especially the possible increased frequency of extreme events. Changing patterns and occurrences of events may result in an expansion of the potentially vulnerable areas. **It is considered important that hazard maps in use by local planning authorities are regularly reviewed and updated to take account of changing conditions.** This may involve:

    11a. the development of coordinated monitoring programmes, involving measurement of erosion and deposition rates;

    11b. maintaining detailed records of erosion, deposition and flooding events;

    11c. modelling the changing relationships between extreme climate events and erosion, deposition and flood events;

    11d. the development of a range of scenarios based on past variations in climate and sea level;

    11e. reviewing regularly the statistical assessment of the return periods of extreme events.

Chapter 12: References

Acreman M.C. 1985. The effects of afforestation on the flood hydrology of the Upper Ettrick Valley. Scottish Forestry 39, 89–99.

Archer D.R. 1987. Improvement in flood estimates using historical information on the R. Wear and Durham. National Hydrology Symposium, Hull.

Arup Economics and Planning 1993. Tests of minerals planning policy options: a model of aggregates supply and demand. Report to the Department of the Environment. HMSO.

BACMI 1992. Statistical yearbook.

Best R.H. 1976. The extent and growth of urban land. The Planner 62, 8–11.

Bevan K. 1993. Riverine flooding in a warmer Britain. Geographical Journal 159, 157–161.

Boardman J. 1993. Climate change and soil erosion in Britain. Geographical Journal 159, 175–183.

Brampton A. 1992. Beaches – the natural way to coastal defence. In M.G. Barrett (ed) Coastal zone planning and management, 221–229. Thomas Telford.

Bray M.J., Carter D.J. and Hooke J.M. 1992. Sea level rise and global warming: scenarios, physical impacts and policies. Report to SCOPAC, Dept. Geography, University of Portsmouth.

Carter R.W.G, 1989. Rising sea level. Geology Today, 5, 63–67.

Department of the Environment 1992a. PPG1 General policy and principles. HMSO.

Department of the Environment 1992b. PPG12 Development plans and regional planning guidance. HMSO.

Doornkamp J.C. (ed) 1990. The greenhouse effect and rising sea levels in the UK. M1 Press.

English Nature 1992. Campaign for a Living Coast. English Nature, Peterborough.

Geomorphological Services Limited 1986–1987. Review of Landsliding in Great Britain. Reports to the Department of the Environment.

Houghton J.T., Jenkins G.J. and Ephraums (eds) 1990. Climate change: The IPCC Scientific Assessment. Cambridge University Press.

Hulme M., Hossell J.E. and Parry M.L. 1993. Future climatic change and land use in the United Kingdom. Geographical Journal, 159, 131–147.

Inglis C. and Allen F., 1957. The regime of the Thames estuary as affected by currents, salinities and river flow. Proceedings of the Institution of Civil Engineers, 7, 827–868.

Innes J.L. 1983. Lichenometric dating of debris flow deposits in the Scottish Highlands. Earth Surface Processes and Landforms, 8, 579–588.

Jager J. and Ferguson H.L. (eds) 1991. Climate change : Science, impacts and policy. Cambridge University Press.

Jones D.K.C. and Lee E.M. 1994. Landsliding in Great Britain. HMSO.

Lamb H.H. 1991. Historic storms of the North Sea, British Isle and Northwest Europe. Cambridge University Press.

Macklin M.G., Rumsby B.T. and Newson M.D. 1991. Historic floods and vertical accretion in fine grained alluvium in the Lower Tyne Valley, North East England. In R.D. Hey (ed) Dynamics of gravel bed rivers. Wiley and Sons Limited.

Ministry of Agriculture, Fisheries and Food 1993. Project appraisal guidance notes for flood and coastal defence.

Neal J. and Parker D.J. 1988. Floodplain encroachment: a case study of Datchet UK. Middlesex University Geography and Planning Paper No. 22.

Newson M. and Lewin J. 1991. Climatic change, river flow extremes and fluvial erosion – scenarios for England and Wales. Progress in Physical Geography 15, 1–17.

NERC 1975. Flood Studies Report. Institute of Hydrology.

NRA 1992. Corporate Plan.

Parry M.L. and others 1991. The potential effects of climate change in the United Kingdom. Department of the Environment. HMSO.

Pethick J. and Burd F. 1993. Coastal defence and the environment: a guide to good practice. MAFF Publications.

Rendel Geotechnics 1995. Review of Erosion, Deposition and Flooding: investigation and management.

Secretary of State and others, 1994. Sustainable Development: the UK strategy. HMSO.

Shennan I., 1989. Holocene crustal movements and sea level changes in Great Britain. Journal of Quaternary Sciences, 4, 77–89.

Shennan I. 1993. Sea level change and the threat of coastal inundation. Geographical Journal 159, 148–156.

Short T., 1749. A general chronological history of the air, weather, seasons, meteors etc., Longman.

Taylor G.G.M. 1970. Ploughing practice in the Forestry Commission. Forest Record No. 73. HMSO.

Thom A.S. and Ledger D.C. 1976. Rainfall, runoff and climatic change. Proceeding of the Institution of Civil Engineers, 61, 633–652.

Woodworth P.L. 1990. A search for accelerations in records of European mean sea level. International Journal of Climatology 10, 129–143.

Woodworth P.L., Shaw S.W. and Blackman D.L. 1991. Secular trends in mean tidal range around the British Isles and along the adjacent European coast. Geophysics Journal International 104, 593–609.

Appendix 1   The Historical Records Database

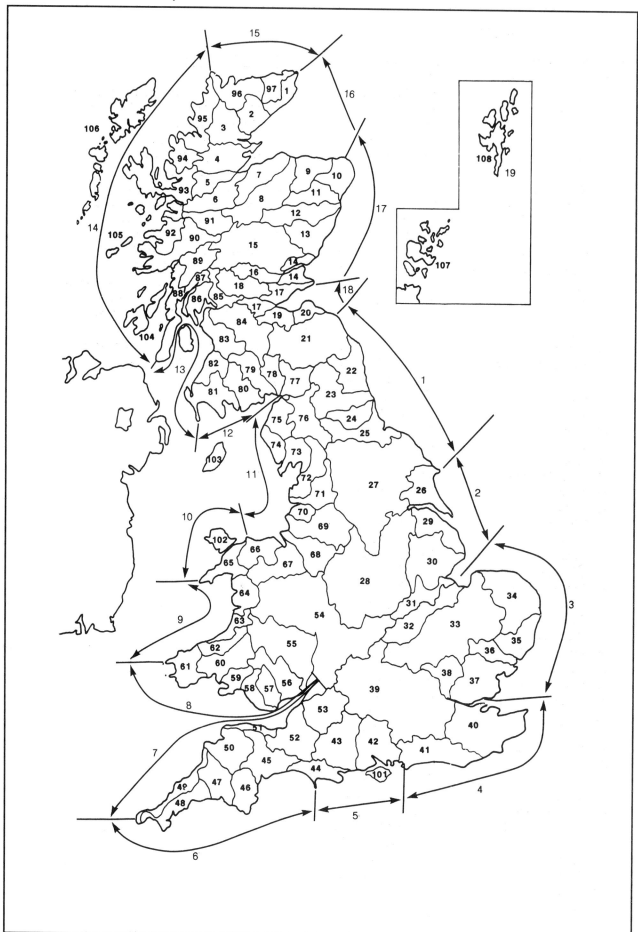

Figure A.1 Catchments and coastal systems.

Key to Hydrometric Areas

1	East Caithness		54	Severn
2	Brora & Helmsdale		55	Wye
3	Stathcarron		56	Usk & Ebbw Vale
4	Cromarty		57	Taff
5	Stathglass		58	Neath
6	Ness		59	Loughour
7	Findhorn		60	Tywi
8	Spey		61	Pembroke
9	North Banffshire		62	Teifi
10	East Aberdeenshire		63	Aeron
11	Don		64	Dyfi
12	Dee (Aberdeenshire)		65	Lleyn
13	South Esk & North Esk		66	Conwy & Clwyd
14	Firth of Tay		67	Dee & Weaver
15	Tay		68	Mersey
16	Earn		69	S. Lancashire
17	Firth of Forth		70	N. Lancashire
18	Forth		71	Ribble
19	Almond		72	Wyre & Lune
20	East Lothian: Tyne		73	South Lakes
21	Tweed		74	West Cumbria
22	Coquet		75	North Lakes
23	Tyne		76	Eden
24	Wear		77	Solway
25	Tees		78	Annan
26	Hull		79	Nith
27	Ouse (Yorkshire)		80	Ken
28	Trent		81	Cree & Luce
29	Ancholme		82	Doon
30	Witham		83	Irvine
31	Welland		84	Clyde
32	Nene		85	Lomond
33	Ouse (Norfolk)		86	Long
34	Norfolk Broads		87	Fyne
35	Blyth		88	Argyllshire
36	Orwell		89	Etive
37	Blackwater		90	Linnhe
38	Lea		91	Garry
39	Thames		92	Sunart
40	Medway, Stour & Rother		93	Wester Ross
41	Cuckmere & Arun		94	Maree
42	Test		95	West Sutherland
43	Stour & Avon		96	North Sutherland
44	South Dorset		97	Thurso
45	Exe			
46	South Devon		102	Anglesey
47	Tamar		103	Isle of Man
48	South Cornwall		104	Islay
49	North Cornwall		105	Inner Hebrides
50	Taw & Torridge		106	Western Isle
51	Exmoor		107	Orkney
52	Parrctt		108	Shetland Isles
53	Avon			

Key to Littoral Cells

1. St Abb's Head to Flamborough Head
2. Flamborough Head to The Wash
3. The Wash to the Thames
4. The Thames to Selsey Bill
5. Selsey Bill to Portland Bill
6. Portland Bill to Land's End
7. Land's End to The Severn
8. The Severn to St David's Head
9. St David's Head to Bardsey Sound
10. Bardsey Sound to Great Orme
11. Great Orme to Solway Firth
12. Solway Firth to Mull of Galloway
13. Mull of Galloway to Mull of Kintyre
14. Mull of Kintyre to Cape Wrath
15. Cape Wrath to Duncansby Head (including Orkneys)
16. Duncansby Head to Kinnard Head
17. Kinnard Head to Fife Head
18. Fife Head to St Abb's Head
19. Shetlands

Table A.1 The distribution of events by catchment.

Catchment No.	Name	Area (km²)	Hillslope	Fluvial	Coastal	Aeolian	Total
0	Unspecified	n/a	24	212	43	–	279
1	East Caithness	840	1	1	1	–	3
2	Brora & Holmedale	1308	1	2	–	–	3
3	Strathcarron	1988	1	6	–	–	7
4	Cromarty	2162	5	14	1	–	20
5	Strathglass	1156	–	9	–	–	9
6	Ness	1947	3	27	1	3	34
7	Findhorn	1807	3	11	2	2	18
8	Spey	2809	3	17	–	–	20
9	North Banffshire	1614	2	4	2	–	8
10	East Aberdeenshire	1485	1	–	2	3	6
11	Don	1355	–	5	–	–	5
12	Dee (Aberdeenshire)	1914	–	12	1	–	13
13	South Esk & North Esk	1885	2	9	4	1	16
14	Firth of Tay	1049	3	6	4	1	14
15	Tay	4985	6	36	–	1	43
16	Earn	890	1	13	–	–	14
17	Firth of Forth	1337	4	13	5	–	22
18	Forth	1604	5	10	1	–	16
19	Almond	909	6	6	1	–	13
20	East Lothian: Tyne	571	1	11	–	–	12
21	Tweed	4579	6	22	–	–	28
22	Cocquet	2788	3	15	–	–	18
23	Tyne	2748	2	39	–	–	41
24	Wear	1236	1	11	–	–	12
25	Tees	2134	3	8	3	–	14
26	Hull	2142	5	7	5	–	17
27	Ouse (Yorkshire)	11457	18	73	14	–	105
28	Trent	10218	9	69	–	–	78
29	Ancholme	1920	1	5	5	–	11
30	Witham	3378	2	16	3	–	21
31	Welland	1671	1	8	–	–	9
32	Nene	2354	4	16	1	–	21
33	Ouse (Norfolk)	8472	5	34	4	4	47
34	Norfolk Broads	3794	6	18	58	2	84
35	Blyth	1649	3	7	7	–	17
36	Orwell	1039	1	2	–	–	3
37	Blackwater	3284	8	21	26	–	55
38	Lea	1381	5	24	–	–	29
39	Thames	10793	27	113	3	–	143
40	Medway, Stour & Rother	4668	12	33	112	–	157
41	Cuckmere & Arun	3264	9	17	24	–	50
42	Test	2770	2	6	4	–	12
43	Stour & Avon	2988	4	12	1	–	17
44	South Dorset	1397	2	7	37	–	46

Table A.1 (cont ...)

Catchment No.	Name	Area (km²)	Hillslope	Fluvial	Coastal	Aeolian	Total
45	Exe	2478	12	50	14	–	76
46	South Devon	1529	2	10	7	–	19
47	Tamar	1842	3	7	–	–	10
48	South Cornwall	1632	–	22	4	–	26
49	North Cornwall	1284	1	5	1	–	7
50	Taw & Torridge	2066	5	13	–	–	18
51	Exmoor	478	2	5	2	–	9
52	Parrett	2783	4	30	3	–	37
53	Avon	2132	9	33	1	–	43
54	Severn	11402	12	69	1	–	82
55	Wye	4197	10	40	–	–	50
56	Usk & Ebbw Vale	1709	24	35	3	–	62
57	Taff	1057	28	44	2	–	74
58	Neath	984	9	27	3	1	40
59	Loughor	1005	9	23	3	–	35
60	Tywi	2100	4	20	1	–	25
61	Pembroke	1524	–	7	4	–	11
62	Teifi	978	2	11	–	–	13
63	Aeron	786	3	4	4	–	11
64	Dyfi	1442	4	14	9	1	28
65	Lleyn	1340	7	10	6	–	23
66	Conwy & Clwyd	1406	3	17	4	–	24
67	Dee & Weaver	2259	5	14	–	–	19
68	Mersey	1777	3	9	1	–	13
69	South Lancashire	2760	6	27	1	–	34
70	North Lancashire	650	4	6	–	–	10
71	Ribble	1491	1	12	2	–	15
72	Wyre & Lune	1750	1	7	7	–	15
73	South Lakes	1166	1	10	2	–	13
74	West Cumbria	977	–	–	–	–	0
75	North Lakes	1204	2	4	–	–	6
76	Eden	2498	1	8	–	–	9
77	Solway	2859	8	23	2	–	33
78	Annan	1011	1	9	1	–	11
80	Ken	1450	2	9	–	–	11
81	Cree & Luce	2148	3	8	1	–	12
82	Doon	1081	6	12	2	–	20
83	Irvine	1653	15	31	7	–	53
84	Clyde	3184	22	66	3	–	91
85	Lomond	973	2	15	–	–	17
86	Long	957	1	7	3	–	11
87	Fyne	724	2	1	–	–	3
88	Argyllshire	975	2	3	1	–	6
89	Etive	1520	1	4	–	–	5
91	Garry	3481	7	9	–	–	16

Table A.1 (cont ...)

Catchment No.	Name	Area (km²)	Hillslope	Fluvial	Coastal	Aeolian	Total
92	Sunart	374	–	–	–	–	0
93	Wester Ross	1761	1	1	1	–	3
94	Marce	1184	1	1	1	–	3
95	West Sutherland	2438	–	–	–	–	0
96	North Sutherland	1847	–	3	–	–	3
97	Thurso	922	1	1	1	–	3
101	Isle of Wight	384	2	3	29	–	34
102	Anglesey	728	1	–	1	–	2
104	Islay	unspec	1	3	6	–	10
105	Inner Hebrides	unspec	1	–	–	–	1
106	Western Isles	2897	–	–	–	1	1
107	Orkney	976	–	1	–	–	1
108	Shetland Isles	1429	–	–	2	–	2
			468	1810	521	20	2819

NOTE: Events are counted no more than once for each system in each catchment

Table A.2 The distribution of events by coastal system (littoral cell).

Coastal cell No.	Name	Coastal	Aeolian	Total
1	St Abb's Head to Flamborough Head	15	–	15
2	Flamborough Head to The Wash	12	–	12
3	The Wash to the Thames	79	2	81
4	The Thames to Selsey Bill	129	–	129
5	Selsey Bill to Portland Bill	35	–	35
6	Portland Bill to Land's End	57	–	57
7	Land's End to the Severn	6	–	6
8	The Severn to St David's Head	12	1	13
9	St David's Head to Bardsey Sound	12	1	13
10	Bardsey Sound to Great Orme	1	–	1
11	Great Orme to Solway Firth	16	–	16
12	Solway Firth to Mull of Galloway	3	–	3
13	Mull of Galloway to Mull of Kintyre	12	–	12
14	Mull of Kintyre to Cape Wrath	5	1	6
15	Cape Wrath to Duncansby Head (including The Orkneys)	1	–	1
16	Duncansby Head to Kinnard Head	4	5	9
17	Kinnard Head to Fife Head	9	2	11
18	Fife Head to St Abb's Head	7	1	8
19	Shetlands	1	–	1
		416	13	429

NOTE: Events are counted no more than once for each system in each cell

Table A.3 The distribution of events by decade from 1700 and by century from 1000–1700.

Dates	Hillslope	Fluvial	Coastal	Aeolian	Total
1990 1993	7	20	7	–	34
1980 1989	19	46	6	–	71
1970 1979	15	45	29	–	89
1960 1969	32	52	28	–	112
1950 1959	31	69	29	–	129
1940 1949	18	48	18	–	84
1930 1939	22	56	33	1	112
1920 1929	23	54	36	–	113
1910 1919	6	34	15	–	55
1900 1909	11	27	24	1	63
1890 1899	11	35	18	1	65
1880 1889	4	26	12	–	42
1870 1879	13	35	14	–	62
1860 1869	7	42	11	1	61
1850 1859	7	35	13	–	55
1840 1849	3	35	18	–	56
1830 1839	1	13	13	–	27
1820 1829	–	21	12	–	33
1810 1819	1	11	6	–	18
1800 1809	–	8	5	–	13
1790 1799	1	26	8	1	36
1780 1789	2	18	3	–	23
1770 1779	–	6	1	–	7
1760 1769	–	6	1	–	7
1750 1759	–	5	1	–	6
1740 1749	–	5	4	–	9
1730 1739	–	5	5	–	10
1720 1729	–	1	–	–	1
1710 1719	–	2	1	–	3
1700 1709	–	1	3	–	4
1600 1699	–	40	11	9	60
1500 1599	–	18	7	1	26
1400 1499	–	7	–	3	10
1300 1399	–	10	2	2	14
1200 1299	–	11	6	–	17
1100 1199	–	1	1	–	2
1000 1099	–	1	–	–	1
	234	875	401	20	1530

NOTE: Events are counted no more than once for each system in each decade

Printed in the United Kingdom for HMSO
Dd301775 12/95 C8 G559 10170